D0991486

HANDBOOK OF SIMPLIFIED COMMERCIAL AND INDUSTRIAL WIRING DESIGN

JOHN D. LENK

Consulting Technical Writer

PRENTICE–HALL, INC., *Englewood Cliffs, N.J. 07632*

Library of Congress Cataloging in Publication Data

Lenk, John D.
  Handbook of simplified commercial and industrial
wiring design.

  Includes index.
  1. Electric wiring—Handbooks, manuals, etc.   I. Title.
TK3271.L448   1984   621.319′24   83–23115
ISBN 0–13–381666–4

Editorial/production supervision and interior design: Ellen Denning
Manufacturing buyer: Anthony Caruso

To my very special wife, Irene. Thank you for your patience!

To Jerry Slawney of Prentice-Hall. My special thanks for helping make my books best-sellers.

To my Magic Lambie for helping me write my books.

© 1984 by Prentice-Hall, Inc., Englewood Cliffs, New Jersey 07632

All rights reserved. No part of this book may be
reproduced, in any form or by any means,
without permission in writing from the publisher.

Printed in the United States of America

10 9 8 7 6 5 4 3 2 1

ISBN   0-13-381666-4

PRENTICE-HALL INTERNATIONAL, INC., *London*
PRENTICE-HALL OF AUSTRALIA PTY. LIMITED, *Sydney*
EDITORA PRENTICE-HALL DO BRASIL, LTDA., *Rio de Janeiro*
PRENTICE-HALL CANADA INC., *Toronto*
PRENTICE-HALL OF INDIA PRIVATE LIMITED, *New Delhi*
PRENTICE-HALL OF JAPAN, INC., *Tokyo*
PRENTICE-HALL OF SOUTHEAST ASIA PTE. LTD., *Singapore*
WHITEHALL BOOKS LIMITED, *Wellington, New Zealand*

CONTENTS

PREFACE

This book can be used as a companion to the author's best-selling *Handbook of Simplified Electrical Wiring Design,* and is written to provide the same type of coverage for commercial and industrial wiring design. However, this book stands alone, and the reader need make no reference to the earlier book. Any information common to both residential and commercial/industrial wiring design is available, in complete form, in both books.

As in the earlier work, this book is a guide or a key to the NEC (*National Electrical Code®*). However, this book makes no attempt to duplicate the NEC in boring detail. Instead, the book is *equipment oriented.* That is, the wiring design examples are based on using off-the-shelf equipment to meet NEC requirements for commercial and industrial wiring, rather than on theoretical solutions. This makes it possible for the reader to put the information to immediate use on practical wiring design problems. Of course, the book also gives the background necessary to understand the features, ratings, and so on, found on off-the-shelf electrical equipment.

The purpose of this book is to tell the journeyman electrician or working contractor how to solve specific commercial and industrial wiring problems. The book assumes that the reader is experienced in making installations planned by others, where all the paperwork problems have been solved, and that the readers now want to do their own planning and problem solving.

The information in this book tells the experienced working person how to do some of the work of a contractor. The book touches on estimating, but avoids the legal and business end of electrical contracting. Instead, the book concentrates on practical design, selection of equipment, and planning of electrical wiring.

The book assumes that the reader is an experienced electrician, so very few mechanical details are included (only where necessary to solve wiring design problems). The book assumes also that the reader is familiar with basic electrical equations, Ohm's law, magnetism, and the like. No basic electrical theory is repeated. However, a summary of basic electrical equations and how they are applied to practical design is included in the final chapter.

The book investigates capabilities and limitations of various wiring system alternatives so that the reader can make logical choices to meet specific design requirements. The book also provides shortcuts and guidelines or "rules of thumb" for those instances where a fast, approximate design answer is sufficient. These shortcuts can be applied directly to the information found in electrical equipment catalogs (which is always a good place to start solving wiring design problems).

Chapter 1 discusses the basics of electrical power distribution (generating station to service entrance, and service entrance to interior electrical outlets and loads). Chapter 1 is included as a starting point for practical design, a review for the experienced reader, and as an introduction for the student reader.

Chapter 2 covers conductors, raceways, wireways, and busways. All necessary tables and equations, plus worked-out examples, are included to calculate conductor, raceway, wireway, and busway sizes, materials, and types.

Chapter 3 describes circuit breakers, safety switches, and fuses commonly used in commercial and industrial wiring. The chapter tells what is available, how the equipment is rated, what the ratings mean, and how the ratings affect design of a commercial or industrial wiring system.

Chapter 4 discusses the distribution of electrical power from the service entrance to electrical outlets and load in a commercial or industrial system. Included are tables, equations, and examples to help in finding correct feeder, branch-circuit, and overcurrent protection sizes, including short-circuit interrupt ratings.

Chapter 5 covers grounding systems in electrical wiring and describes the how and why of grounding from a design standpoint, with a discussion of NEC grounding requirements. The chapter also discusses ground-fault circuit interrupters and associated wiring problems.

Chapter 6 describes how transformers can be used to provide the different voltages often required in present-day commercial and industrial wiring systems.

Chapter 7 discusses the design of electrical lighting from the standpoint of providing a given amount of light for a given area, and of selecting the proper control devices for commercial and industrial lighting systems. The chapter includes shortcut calculations to determine how many lamps are required to provide a given illumination level in a given interior space, and how the lamps should be arranged to provide uniform illumination throughout the space.

Chapter 8 discusses the fundamentals of electrical motor control, including how to select starters and controllers of the proper size and rating for a given motor. Also included is information on understanding motor control wiring diagrams.

Chapter 9 is a review, or brief summary, of basic electrical data most needed for the planning and design of commercial and industrial wiring systems. Basic electrical theory is not included, nor is mathematical analysis. Mathematics is used only where absolutely necessary, and then in the simplest form.

Since the book does not require advanced math or theoretical study, it is ideal for the working electrician or contractor who needs a ready reference source for basic equations, calculation techniques, and related information. The book is suited also for schools that teach electrical power distribution analysis with a strong emphasis on practical, simplified design.

Many professionals have contributed their talent and knowledge to the preparation of this book. The author gratefully acknowledges that the tremendous effort required to make this book such a comprehensive work is impossible for one person, and he wishes to thank all who have contributed directly and indirectly. Special thanks are due to the following: W. S. Cahill of General Electric Company; Paul Sackenheim of the IC Division, and Nord Froehlich and Vicki Snapp of the Distribution Equipment Division of Square D Company; Dick Nelson of the Department of Water and Power for the City of Los Angeles; Kenneth Gudger of the Southern California Edison Company; the Illuminating Engineering Society of North America; and Dennis Berry of the National Fire Protection Association.

The author extends his gratitude to Greg Burnell, Dave Boelio, Hank Kennedy, John Davis, Jerry Slawney, Art Rittenberg, Matt Fox, Dave Ungerer, and Don Schaefer of Prentice-Hall. Their faith in the author has given him encouragement, and their editorial/marketing expertise has made many of the author's books best-sellers. The author also wishes to thank Mr. Joseph A. Labok of Los Angeles Valley College for his help and encouragement.

JOHN D. LENK

1

POWER DISTRIBUTION TO COMMERCIAL AND INDUSTRIAL INSTALLATIONS

In this chapter we cover the basics of electrical power distribution from the generating station to the service entrance of the individual commercial or industrial customer, and from the service entrance to the electrical outlets and other equipment operated by electrical power. We also discuss the basic elements of electrical wiring from a simplified, practical standpoint. The systems and equipment that we treat are typical of those in most general use throughout commercial and industrial installations.

We do not dwell on basic electrical theory in any detail, although a summary of electrical theory in Chapter 9 includes such standard subjects as Ohm's law, the calculation of power in single- and three-phase circuits, the power factor, power factor correction, and related subjects necessary for practical wiring design.

No book on electrical wiring is complete without reference to the *National Electrical Code®* (*NEC*)®. Although frequent reference to the Code is made throughout this book, we made no attempt to duplicate the *NEC®* or even to cover every paragraph in full detail, as copies of the Code are available from most utility companies. *Where practical, Code references are made in italic type throughout this book to assist the reader in becoming quickly familiar with the Code's provisions as they apply to simplified wiring design.*

National Electrical Code® and NEC® are Registered Trademarks of the National Fire Protection Association, Inc., Quincy, MA.

1

1-1 *NATIONAL ELECTRICAL CODE*®

To assure that electrical equipment is standardized and properly installed, the National Fire Protection Association (NFPA) has developed a set of minimum standards for electrical installations in homes, stores, industrial plants, offices, and so on. The NFPA also determines manufacturing standards for many electrical devices. The standards, which apply primarily to size, assure that lamps of a given type fit all sockets of the corresponding type, plugs fit receptacles, and so on. The NFPA standards are not to be confused with those of the Underwriters' Laboratories (UL). The UL is a nationally accepted organization that tests all types of wiring materials and devices to make certain that they meet minimum standards for safety and quality.

The NFPA standards concerning installation are incorporated into a book called the *National Electrical Code*® (NEC). Each section of the NEC is supervised by a panel of experts in various areas. These experts check the code against the latest developments in the electrical industry, revising the code as required to reflect changes in modern electrical wiring and equipment. Typically, the NEC is revised every three years.

Although the NEC is a nationally accepted document, all provisions are subject to local interpretation. Each state, county, city, and township may have a code or regulations for installation of electrical equipment or wiring. In some cases the local authority accepts part of the NEC without change, and then publishes regulations that supersede other parts of the NEC. A classic example is the use of conduit that houses electrical wiring (see Chapter 2). Many local authorities require either rigid conduit or armored cable for all wiring (both new work and additions). Regulations in other areas permit nonmetallic (typically plastic) covering for cable used in additions, remodeling, and the like. In still other areas, nonmetallic cable can be used for homes and other residential installations, but not for commercial and industrial work.

Another problem is that local wiring inspectors invariably differ in their interpretation of the provisions of the NEC. For these reasons, the NEC provisions should be considered to be *minimum requirements* for any electrical wiring installation. Also, the NEC's provisions apply in the absence of a specific requirement by the local authority.

The NEC should not be considered to be an instruction manual. However, many NEC standards form the basis for the solution of problems related to the design of electrical wiring. For example, the NEC permits a maximum voltage drop of 5% (of the source voltage) from the service entrance (point where electrical power enters the building or industrial plant) to the most distant electrical outlet. This voltage drop limit of 5% sets the size of electrical wiring in each circuit throughout the system. (Voltage drop and wire size are discussed fully in Chapter 2.)

To sum up, anyone involved in the design of electrical wiring should have a copy of the NEC (of the latest issue), plus any local codes or regulations.

If there is doubt as to which local agency governs the installation of electrical systems, contact the local utility company. In fact, the local utility company should always be contacted in regard to electrical wiring design problems, since each utility company has its own requirements for electrical service (size and type of service entrance equipment, length of service cable, etc.).

Keep in mind that any references made to the NEC in this book are subject to change, since the NEC is in a constant state of revision. Always consult the latest issue of the NEC as the final authority for wiring design.

1-2 ELECTRICAL POWER DISTRIBUTION TO THE CUSTOMER

The distribution of electrical power from the generating station to the individual customer is shown in Fig. 1-1. Note that this system is "typical." That is, the electrical power is alternating current at 60 hertz (Hz) and is made through three-phase transmission lines. Of course, the distribution of electrical power is not the same for all areas of the country. In addition, the distribution system shown in Fig. 1-1 is the responsibility of the utility ty company, not of the contractor or electrician. However, the distribution of electrical power to the service entrance is the starting point for the solution of any electrical wiring design problem.

As shown in Fig. 1-1, the generating station (hydroelectric, steam, nuclear, etc.) produces voltages in the range 12 to 24 kilovolts (kV). This voltage is stepped up (by transformers, as described in Chapter 6) to transmission voltages in the range 60 to 110 kV and higher. The reason for such high transmission voltages is because the voltage drops as the electrical power passes along the transmission cables from generating station to substation. As a guideline, the transmission voltage must be approximately 1000 V for each mile of transmission.

Power is distributed from the substation to individual customers (or groups of customers) through three-phase transmission lines. Typically, the voltage across each phase is 2400 V. Since the transformers at the substation output are connected in three-phase wye configuration, the voltage from phase to phase is 4160 V. (Readers not familiar with three-phase electrical power should refer to Chapters 6 and 9.) The 2400/4160-V power is stepped down by transformers to voltages suitable for each customer's needs. Typical distribution systems are described in the following sections.

1-2.1 120/240-V Single-Phase Distribution

Most lighting and appliances in offices and small buildings require 120 or 240 V. Generally, 240 V is used for larger appliances such as office or plant air conditioning. Note that most appliances operate on any voltage between 110

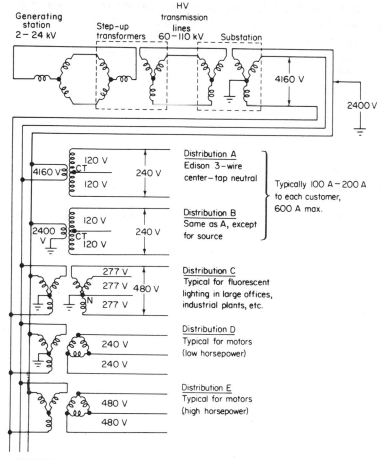

FIGURE 1-1 Simplified diagram of electrical power distribution from generating station to individual customer.

and 120 V (or 220 and 240 V). The actual power supplied by most utility companies is at voltages within these ranges. Often, such power is referred to as 110, 115, or 120 V, depending on locality. Multiples are referred to as 220, 230, and 240 V. In this book, we standardize on 115 or 120 V (and 230 or 240 V). However, the terms 110 and 220 still exist, and are in common use.

Under normal conditions, distribution circuits are operated by the utility company so as to maintain voltage levels to customers within certain ranges. The voltage levels shown in Fig. 1-2 are those supplied by the Southern California Edison Company and are typical of most utility companies.

The 120/240-V system shown as distribution A in Fig. 1-1 is supplied by a three-wire system. The transformer primary is fed by 4160 V, taken across two lines of the three-phase transmission. This system, sometimes referred to

Service voltages

Nominal two-wire and multiwire service voltage	Minimum voltages to all services	Maximum service voltages on residential and commercial distribution circuits	Maximum service voltage on agricultural and industrial distribution circuits
120	114	120	126
208	197	208	218
240	228	240	252
277	263	277	291
480	456	480	504

Customer utilization voltages

Nominal utilization voltage	Customer utilization voltages for fully satisfactory performance	
	Minimum	Maximum
120	110	125
208	191	216
240	220	250
277	254	289
480	440	500

FIGURE 1-2 Typical minimum and maximum service voltages, and customer utilization voltages.

as the *Edison three-wire system,* is typical for at least 90% of the offices, stores, and other small buildings in the United States. (The theory of the three-wire system is discussed in Chapter 9; practical connections and applications are discussed throughout other chapters.)

For the purposes of this chapter, it is sufficient to say that there is 120 V from the center tap to either end of the transformer secondary, and 240 V between the ends of the secondary. The three wires (or service conductors, usually called the *service drop*) are brought into the customer's entrance. However, not all customers use three wires, particularly in older commercial and industrial applications. In such installations, where there are no major electrical appliances, 240 is not needed, so only two wires are used beyond the service entrance.

The center-tap wire (known as the *neutral* or *ground wire*) is always used. The center-tap wire forms a complete circuit with either of the end taps to provide 120 V.

The 120/240-V system shown as distribution B in Fig. 1-1 is identical to distribution A in that the customer is provided with a 120/240-V three-wire Edison system at the service entrance. However, in this case, the utility company taps across one phase of the three-phase transmission line. Thus the transformer primary is fed by 2400 V rather than by 4160 V. Distribution B makes it possible to use smaller transformers but provides less power. The system of distribution B is generally used when the distance between transformer and service entrance is less than 600 ft, and no more than six 100-A service entrances are served. However, this is the concern of the utility company and not of the wiring designer.

1-2.2   277/480-V Three-Phase Distribution

The 277/480-V system shown as distribution C in Fig. 1–1 is supplied by a four-wire system. The transformer primary (connected in wye configuration) is fed by all three phases of the 2400/4160-V transmission. The transformer secondary (also connected in wye) delivers 277 V across each phase and 480 V between phases. The center tap of all three-phase windings is grounded and requires a fourth neutral (or ground) conductor to the service entrance.

The 277/480-V distribution (also referred to as 265/460-V power) is used primarily for the lighting of large industrial plants and office buildings. Many fluorescent lighting systems are designed for use with 277/480-V power.

The 277/480-V distribution has the disadvantage of requiring four conductors rather than three. However, the higher voltage (277 V versus 120 V) permits longer runs for conductors, or lower currents, or a combination. (The advantages and limitations of 277/480-V power are discussed throughout several chapters.) The main disadvantage (in addition to requiring four conductors) is that most customers also require 120 V for conventional (nonfluorescent) lighting and other applications. Utility companies generally prefer (and provide) reduced rates when they supply only one type of distribution. This means that customers must supply transformers (at their own expense) to convert the 277/480-V three-phase to a 120-V single-phase distribution (or possibly to a 120/240-V three-wire Edison system).

1-2.3   240-V and 480-V Three-Phase Distributions

The 240-V and 480-V systems shown as distributions D and E, respectively, in Fig. 1–1 are supplied by three-wire systems. The transformer primary (connected in wye) is fed by all three phases of the 2400/4160-V transmission. The transformer secondary (connected in delta) delivers 240 V (or 480 V) across each phase. No ground or neutral is required; thus only three service conductors are needed.

The 240-V and 480-V three-phase systems are used primarily for electric motors or similar applications. Single-phase electric motors are impractical for any application except the fractional-horsepower type. Motors used in industry, large buildings, and so on, are operated most efficiently when three-phase (also known as *polyphase*) is used at voltages of 240 V and 480 V (in some cases, 600 V).

When the only supply from the utility company is 240 or 480 V, the customer must provide a transformer to convert the 240-V or 480-V three-phase to 120-V single-phase for conventional lighting and other applications.

1-2.4   High-Voltage Distribution

In addition to the distributions shown in Fig. 1-1, most utility companies also supply higher voltages for special applications. These voltages are typically on the order of 2400, 4160, 4800, 7200, 12,000, 16,500, and 34,500 V. However, high voltages are the exception rather than the rule, so they are not covered in this book.

1-3   THE COMMERCIAL/INDUSTRIAL SERVICE ENTRANCE

Although there is an infinite variety of service entrances and equipment (for each type of customer and at each locality), all service entrances have certain elements in common. Any service entrance must provide the necessary conductors, watthour meters, main disconnect means, overcurrent devices (fuses or circuit breakers), and a system ground.

Each utility company has its own requirements as to service entrance equipment. Generally, the utility company supplies (on request) full details of these requirements. Design and construction of service entrance equipment and systems is not discussed in full. However, the following sections summarize service entrance requirements, particularly as related to wiring design.

1-3.1   Types of Service Entrances

The two basic types of service entrances are overhead and underground. Although the overhead entrance is still used in the majority of commercial and industrial installations, the underground service entrance is being used more and more frequently. The requirements for both types of service entrance are set by the utility company. Although underground service entrances are sometimes installed by the utility company, in the case of an overhead installation, the utility company generally secures the service drop, but leaves installation of the service entrance equipment to the customer.

1-3.2   Service Drop

The service drop consists of those conductors that connect the utility company's transmission lines to the customer service entrance equipment. Figure 1-3 shows some typical requirements for overhead commercial and industrial service drops. Figure 1-4 shows the installation of service drops at various

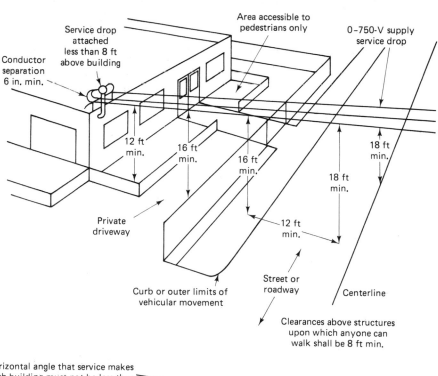

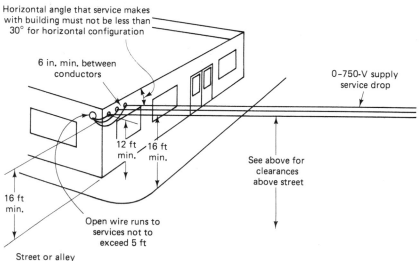

FIGURE 1-3 Typical requirements for overhead commercial/industrial service drops.

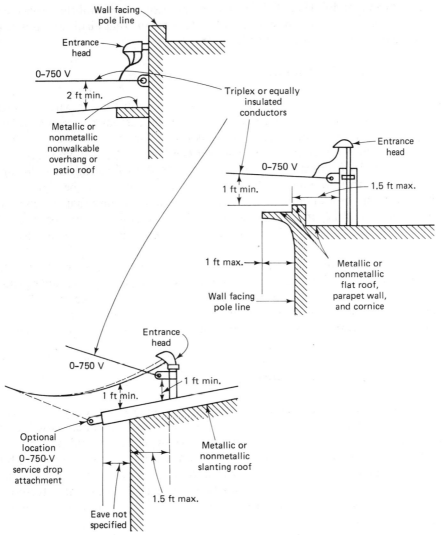

FIGURE 1-4   Installation of service drops at various types of buildings on commercial or industrial premises.

types of buildings on commercial or industrial premises. Keep in mind that these requirements are for one particular utility company, and apply only to overhead service drops. There are corresponding requirements for other types of service drops.

Although the designers of electrical wiring installations need not design the service entrance equipment itself, they must follow the utility company requirements and, in doing so, many of the customer's wiring design requirements are set. For example, to meet the utility company requirement of a service drop location (Fig. 1-3), it may be necessary to install the service entrance at only one location in a building. This means that all electrical wiring must be designed to start at that location.

1-3.3   Service Conductors

Figures 1-3 and 1-4 show that there are three conductors or wires running from the utility company transmission line to the customer's premises. This is typical for most commercial and industrial service drops, except in those rare cases where three-phase four-wire service (Sec. 1-2.2) is involved.

The utility overhead wires are connected to the service entrance through the *entrance head.* In any installation, the entrance head must be higher than the highest or top insulator. Some utility companies require that the conductors must extend beyond the entrance head by a certain amount (such as 36 in.) and be positioned to form loops that prevent water from entering the service.

Often, the three service conductors are colored as follows: white (for the neutral or ground wire), black, and red (or possibly orange). Wire and conduit sizes, as well as current capacities, are discussed fully in Chapter 2. Typical service entrance wiring is covered in Chapter 4. The three conductors pass through the entrance head and service conduit (or raceway) to the service entrance box.

1-3.4   Service Entrance box

Some form of service entrance box is included in all types of service entrances. The service entrance box is also known as the fuse box, main box, switch box, pull box, circuit breaker box, meter box, the switchboard, and possibly the panelboard (although a true panelboard is not used as a service entrance, but within the building between the *feeders* and *branch circuits;* see Sec. 1-3.10).

There are many service entrance box configurations available for commercial and industrial wiring installations. Figures 1-5 through 1-8 show a few typical examples.

Master Meter with Main Disconnect.   Figure 1-5 shows a typical commercial/industrial service entrance box arrangement where all of the loads are metered on one master meter, and where a main service disconnect (also known as the *main* or *service main*) is required by local code. In Fig. 1-5 the service conductors are connected to the meter before the conductors reach the main disconnect. (The main disconnect can be a pull-type disconnect, a switch, or a circuit breaker.) No matter what device is used, when the main disconnect is pulled or opened, all power is removed from the load side of the disconnect.

The conductors at the load side of the main disconnect pass to a number

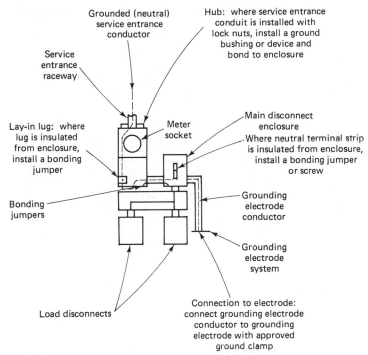

FIGURE 1-5 Single-meter commercial/industrial service with a neutral, where a main disconnect means is required.

of load disconnects (probably circuit breakers, but possibly fused switches). Typically, one load disconnect is provided for each office, store, or other load subdivision. A *panelboard* (sometimes called a circuit breaker board) located within the office or store provides additional control and overcurrent protection.

Individual Meters without Main Disconnect.　Figure 1-6 shows a commercial service entrance where each of the loads is metered individually, and where a main service disconnect is not required by local code. Generally, this is where there are six subdivisions of power (six offices, six stores, etc.) or less. In Fig. 1-6, the service conductors are connected to individual meters before the conductors reach the individual service switches, located in enclosures below the meters. (Although called "service switches," the individual disconnects can be circuit breakers.)

With the arrangement of Fig. 1-6, it is necessary to pull or open individual disconnects to remove power to the corresponding store, office, or other subdivision of power. The conductors pass from the individual disconnects to panelboards within the offices or stores.

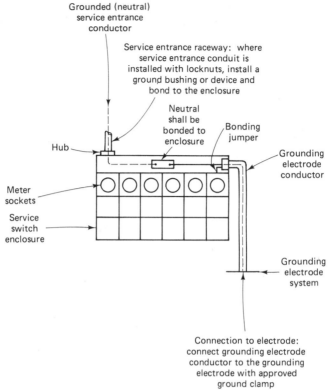

FIGURE 1-6 Typical commercial service entrance where each of the loads is metered individually, and where a main service disconnect is not required by local code.

Individual Meters with Main Disconnect.   Figure 1–7 shows a typical commercial service entrance where each of the loads is metered individually and where a main service disconnect is required by local code. Generally, this is where there are more than six subdivisions of power. In Fig. 1–7, the service conductors are connected to the main disconnect before the conductors reach the individual meters. The conductors then pass to the individual offices and stores through meter switches (or circuit breakers) located in enclosures below the meters.

With the arrangement of Fig. 1–7, it is possible to remove power to the office or store when the corresponding individual disconnect is pulled or opened. However, it is also possible to remove power to all subdivisions simultaneously by pulling or opening the main service disconnect.

Service Entrance Capacity.   The service box arrangements shown in Figs. 1–5 through 1–7 are generally used when single-phase service is 200 A or less. Figures 1–5 through 1–7 can also be used (by most utility companies)

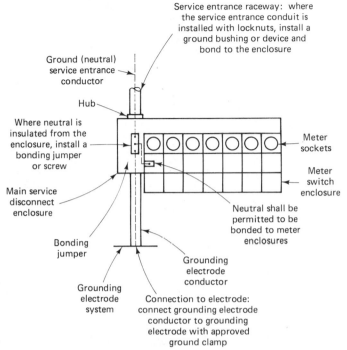

FIGURE 1-7 Typical commercial service entrance where each of the loads is metered individually, and where a main service disconnect is required by local code.

when the service is between 200 and 400 A. However, with such higher currents, some form of *switchboard* is usually recommended. Most utility companies require a switchboard when the service is above 400 A.

Also, most utility companies do not permit single-phase service installations with capacities in excess of 600 A. Similarly, with the higher currents, *bus bars* are used instead of wire conductors, as discussed in Chapter 2. Again, this is at the option of local utilities and local code.

Switchboard Service Entrance.   Figure 1-8 shows a switchboard service entrance, which is typical for industrial installations and many larger commercial applications, such as large shopping centers and high-rise office buildings. The switchboard contains a main disconnect, as well as individual disconnects and meters. Generally, a master meter is used. However, there may be more than one meter in some applications. Also, since higher currents (over 200 A) are involved, *current transformers* are generally used with the meters to reduce the meter size, as discussed in Sec. 1-3.8 and Chapter 6). The current transformers are located in the same compartments as the meters.

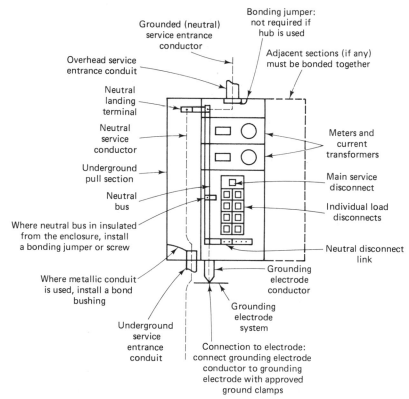

FIGURE 1-8   Typical switchboard service entrance.

1-3.5   Grounding at the Service Entrance

Note that Figs. 1–5 through 1–8 all show a grounding system. The subject of grounding is discussed fully in Chapter 5. For the purpose of this chapter, it is sufficient to say that the neutral or ground wire of any service entrance must be grounded at the service entrance.

1-3.6   Overcurrent Protection

There are two basic types of overcurrent protection in common use, fuses and circuit breakers. These devices protect circuits and wires against overloads and short circuits. A third type of overcurrent device, known as a *ground-fault circuit interrupter* (GFCI) or simply *ground-fault interrupter,* detects small amounts of current leakage and interrupts the circuit when the leakage reaches a certain level. A complete discussion of GFCIs is presented in Chapter 5. In this chapter we concentrate on conventional fuses and circuit breakers used at the service entrance.

Fuses.   Fuses are self-destructive overcurrent devices. That is, fuses are destroyed when interrupting the circuit during an overcurrent condition. Fuses are made of metal with a low melting temperature. The amount of metal is calibrated so that the fuse metal melts at a specific current rating. Fuses are connected in series with the load so that the circuit is opened when the metal melts. In some switchboards (used in heavy-duty industrial applications) fuses are combined with switches, resulting in a device known as a *fusible switch*. Both fuses and switches are discussed in Chapter 3.

Circuit Breakers.   Circuit breakers (also known simply as breakers) are reusable overcurrent devices. Breakers interrupt a circuit under overcurrent conditions and can then be reset. Some breakers operate by *magnetic means,* whereas other breakers are *thermally operated.* Most breakers open the circuit within a fraction of a second during overcurrent. There are circuit breakers (usually of the thermal type) that are "tripped" (open the circuit) after a specific time interval.

Circuit breakers are generally available with one, two, or three poles, all mounted on the same assembly. This permits up to three lines or conductors to be interrupted simultaneously. As discussed in Chapter 5, the ground or neutral wire of any service is *never* provided with a circuit breaker or fuse.

Circuit breakers are available in a wide range of capacities (typically 15 to 6000 A). Several manufacturers provide circuit breakers of different capacities that can be inserted into preassembled enclosures. A main or master circuit breaker may be included in the same enclosure. The enclosures are usually provided with extra space for additional circuit breakers (to be added at a future time as load requirements increase). Circuit breakers are discussed in Chapter 3.

1-3.7 Short-Circuit Withstand Ratings

Both fuses and circuit breakers protect against a short circuit by opening the circuit when *continuous* currents exceed the rated value. A 200-A breaker used as a main disconnect opens when a continuous current over 200 A passes through the service entrance conductors. However, circuit breakers (and switches) must be capable of withstanding *instantaneous currents* (sometimes known as *fault currents*) far in excess of the continuous current ratings.

As an example, the same 200-A breaker might be subjected to 10,000 A for a brief time interval before the breaker opens, and must therefore have a *withstand rating* of 10,000 A or greater. Such high currents can burn out circuit breakers and switches or (generally worse) weld the contact together. The ability to interrupt or open the circuit under short-circuit conditions is referred to as the *interrupt rating* or the *AIC rating,* as discussed in Chapter 3. When a circuit breaker or switch is subjected to currents greater than the withstand or interrupt ratings, the device is generally destroyed, possibly destroying the conductors and other equipment.

The instantaneous short-circuit or fault currents depend (primarily) on voltage and conductor resistance. For a given voltage, a shorter conductor has less resistance, and thus higher possible short-circuit current. Since conductors at the service entrance are those of the utility company, the withstand ratings of the service entrance disconnects must be based on the utility company's contribution to the short-circuit currents. For example, a direct short at the load side of the main disconnect produces short-circuit currents based on service voltage and service conductor resistance. If the short occurs further *downstream* (between the service and some point within the building), the resistance of the conductors (between the service and the short) must be added to the service conductor resistance to find the current.

Short-circuit withstand ratings are generally no problem in residential work, and small commercial wiring installations, since such service involves continuous currents of 200 A or less, and short-circuit currents of 10,000 A or less. Circuit breakers capable of handling 10,000-A short-circuit currents are readily available, particularly those breakers designed for main disconnect duty.

Short-circuit withstand ratings are generally a problem when the service exceeds 200 A, and is therefore of concern in larger commercial/industrial wiring design. Figure 1–9 shows some typical fault current values. The values shown are the utility company's *maximum contribution* to fault current, and can be used to estimate necessary withstand ratings for service entrance circuit breakers (and switches). For example, a 600-A (or less) circuit breaker used as a main disconnect at a service entrance must be capable of withstanding at least 42,000-A fault currents. Again, the values shown in Fig. 1–9 apply to only one utility company (Southern California Edison Company), but are typical. The subject of withstand and interrupt ratings is discussed further in Chapter 3.

Phase	Service voltage	Service entrance ampacity	Utility company's contribution to fault currents will not exceed:
Single-phase	120/240	600 A or less	42,000 A
Three-phase	120/208 or 240	2000 A or less	42,000 A
Three-phase	480	1200 A or less	30,000 A

FIGURE 1–9 Typical values of maximum contribution to fault (short-circuit) currents at the service entrance by the utility company.

1-3.8   Metering

All service entrances include at least one meter to record in kilowatthours the energy consumed by the load. (Kilowatthours and metering are discussed in Chapter 9.) Figure 1–10 shows the basic connections for typical metering. For single-phase 120/240-V service, the meter has two current coils and one voltage coil. For three-phase three-wire service, two voltage coils and two current coils are required. For three-phase four-wire service, three current coils and three voltage coils are needed.

Most kilowatthour meters are plugged into meter sockets. Figure 1–11 shows typical meter socket jaw arrangements for various services. Note that some of the arrangements are for *totalized metering*. Some utility companies offer reduced rates when all service in one building can be totalized on one meter. In some cases it is more economical or practical for the customer to receive more than one type of service from the utility company. The utility company may require more than one meter, or totalized meters, when more than one type of service is provided, or when the current is beyond the capacity of a single meter. (As discussed in Chapter 6, the alternative to multiple service is for the customer to use transformers and convert one type of service to other services.)

When services are rated over 200 A, *current transformers* are used with the kilowatthour meters to reduce the physical size of the meter. Also, where high voltages (generally above 600 V) are involved, *voltage or potential transformers* are used to reduce the voltages to a lower value within the range of the meter. Such transformers are known as *instrument transformers* and are discussed in Chapter 6.

1-3.9   Power Factor Correction

Utility companies generally require that the power factor for all types of service be at some value not less than about 0.8 to 0.85. Some utility companies add a surcharge for power factors below 0.8. Power-factor correction is usually done by the addition of capacitors located at the service entrance or by means of synchronous motors. Power factor and power-factor correction are discussed in Chapter 9.

1-3.10   Feeders, Subfeeders, Branch Circuits, and Panelboards

Electrical power distribution from the service entrance is made by feeder and branch circuits (and possibly subfeeders and panelboards) to the ultimate load. All of these elements are discussed fully in Chapter 4. The following is a summary.

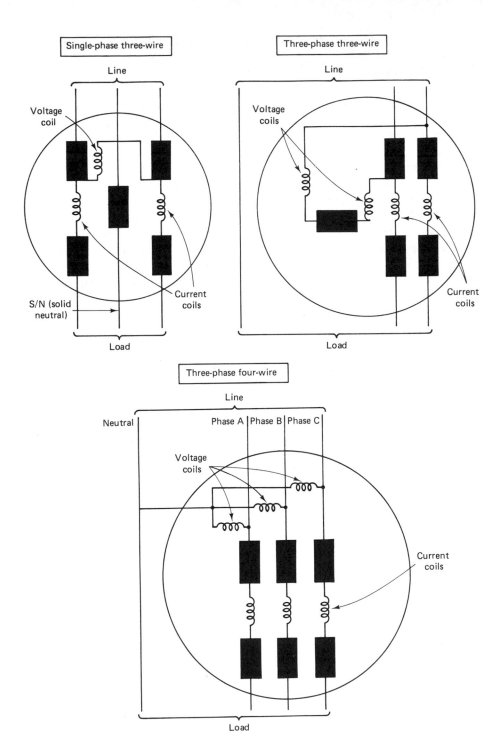

FIGURE 1-10 Basic connections for kilowatthour metering.

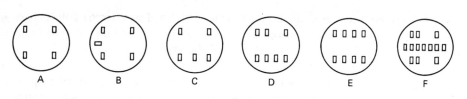

Type of service	Self-contained terminals	With current transformer terminals
Single-phase 120-V two-wire	4: Fig. A	—
Single-phase 480-V two-wire	4: Fig. A	5: Fig. B
Single-phase 240/480-V three-wire	4: Fig. A	—
Single-phase 277/480-V three-wire	5: Fig. C	—
Single-phase 240-V three-wire	4: Fig. A	5: Fig. B
Single-phase 120/208-V three-wire	5: Fig. B or C	—
Three-phase 120/208-V four-wire	7: Fig. D	13: Fig. F
Three-phase 240-V[a]	5: Fig. C	8: Fig. E
Three-phase 480-V[b]	5: Fig. C	8: Fig. E
Three-phase 120/240-V four-wire delta[c]	7: Fig. D	—
Three-phase 120/240-V four-wire delta, above 200 A[c]	—	13: Fig. F
Three-phase 277/480-V four-wire	7: Fig. D	13: Fig. F
Three-phase 2400-, 4800-, or 34,500-V three-wire	—	8: Fig. E
Three-phase 4160-V four-wire	—	13: Fig. F
Three-phase 12,500-V three-wire	—	8: Fig. E
Three-phase 16,500-V three-wire	—	8: Fig. E
Totalizing two three-wire single-phase circuits	—	8: Fig. E
Totalizing three three-wire single-phase circuits	—	13: Fig. F

[a] Single-phase grounded fourth-wire run for grounding only.

[b] 480-V three-wire service may be supplied in some areas.

[c] Mid-tap grounded fourth-wire run for a neutral and/or for grounding.

FIGURE 1-11 Typical meter socket jaw arrangements for various services.

Branch circuit.   A branch circuit can be defined as a set of conductors that extend beyond the last overcurrent device in the system. Typically, a branch circuit supplies only a small part of the system.

Feeder.   A feeder is a set of conductors that supply a group of branch circuits. In a simple distribution system such as that found in most single-family homes, there are no feeders. The branch circuits are supplied directly from the service entrance, through overcurrent devices (fuses and circuit breakers) located at the service entrance. In most commercial and industrial systems, the branch circuits are supplied through feeders. The feeders receive their power through overcurrent devices at the service entrance, and deliver power to the branch circuits through overcurrent devices on panelboards (also known as circuit breaker boards, load centers, etc.). The panelboards are located throughout the building.

Panelboards.   Panelboards are assemblies of overcurrent devices contained in a cabinet or panel accessible only from the front. The overcurrent devices can be either for feeders or branch circuits.

Subfeeders.   When many feeders are necessary, a second panelboard is installed at some point within the building away from the service entrance. This panelboard is supplied with a feeder from the service entrance, and supplies the various branch-circuit panelboards through subfeeders.

1-3.11   Basic Branch-Circuit Wiring

Figures 1–12 and 1–13 show the wiring for portions of typical branch-circuit panelboards. Figure 1–12 shows the wiring for a 120/240-V single-phase three-wire system. There are two basic concerns in design (aside from the current capacities described in Chapter 4).

The system of Fig. 1–12 should be balanced. That is, both of the 120-V distributions should have *approximately the same load*. Most utility companies require that the loads be ''reasonably balanced'' on either side of the neutral conductor. In Fig. 1–12, each 120-V distribution supplies four branch circuits.

The 120-V distribution requires only a single-pole overcurrent device (circuit breaker in this case) for each branch circuit. However, the 240-V distribution requires a double-pole circuit breaker. *Note that the neutral conductor is never provided with an overcurrent device in any distribution.*

Figure 1–13 shows the wiring for a 277/480-V three-phase four-wire system. This system is used mostly for lighting and three-phase motors. The basic concern here is that there be a *reasonable balance between each phase* of the three phases. In Fig. 1–13, each 277-V phase supplies three branch circuits.

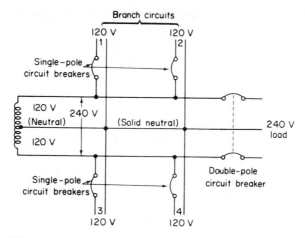

120 V branch circuit loads 1 and 2 = loads 3 and 4 (approx)

FIGURE 1-12 Typical 120/240-V single-phase three-wire system.

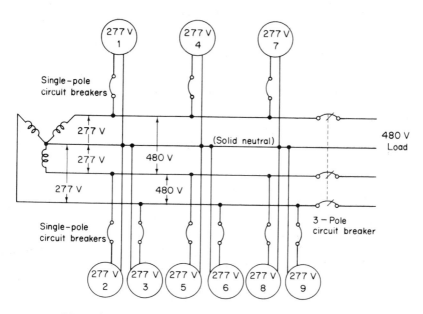

277 V branch circuit loads 1,4 and 7 = 2,5,8 = 3,6,9 (Approx)

FIGURE 1-13 Typical 277/480-V three-phase four-wire system.

Each 277-V branch circuit requires only a single-pole overcurrent device. However, the three-phase 480-V distribution requires a three-pole overcurrent device. Again, the neutral has no overcurrent device. Both Figs. 1–12 and 1–13 show what are called *solid neutral* (S/N) distribution systems.

1-3.12   *More than One Service Entrance*

It may be necessary to have more than one service entrance for larger commercial and industrial wiring installations. This is because most utility companies have some maximum current limits for each service. Obviously, this must be considered in the design of wiring for larger installations. Although each utility company may have different limits, the following requirements of the Southern California Edison Company are typical.

When single-phase is involved, the maximum rating of an individual meter switch for 120/208-V three-wire service is 125 A. Normally, for 120/240- or 240-V service, the maximum rating is 400 A, and 200 A for 480-V services. Under certain operating conditions, permission may be granted for installation of 600-A service equipment for an individual 120/240- or 240-V load. Otherwise, two separate 400-A service installations, or three such installations of 400 A or more, may be required, and totalized meters will normally be available.

When three-phase is involved, the maximum capacity for 120/208, 240, or 277/480 V is 4000 A of connected load. When 4000 A and above is required, two or more services may be installed with totalized metering. The capacities of totalized services should be within 1000 A of each other. Services with combined capacities of less than 4000 A should not be totalized. Again, always consult the local utility company for their electrical service requirements.

2

CONDUCTORS, RACEWAYS, WIREWAYS, AND BUSWAYS

In Chapter 1 we discussed how electrical power is transmitted from the generating station to the service entrance of the individual customer by the utility company. In this chapter we discuss the conductors, raceways, wireways, and busways required for the distribution of electrical power throughout the customer's location. It is obvious that such electrical power must be carried by conductors of suitable size and type. Not so obvious, but equally important, is the fact that the conductors must be installed in suitable raceways. We start by reviewing conductor basics.

2-1 CONDUCTOR BASICS

Most wires (and bus bars) used in present-day commercial and industrial wiring are made of copper (Cu) or aluminum (A1). Both of these materials have a certain resistance or resistivity which must be considered in wiring design. All metals have some resistance to the flow of electrical current, and all metals have a certain conductance. Resistance is the reciprocal of conductance. A metal with low resistance has high conductance, and vice versa.

2-1.1 Units of Resistance and Conductance

The unit of resistance is the *ohm* (Ω), and the unit of conductance is the *mho* (ohm spelled backwards). One ampere of current flows when there is 1 V across 1 Ω.

The *resistivity* of a conductor is defined as the resistance (in ohms) of a specific example of a given cross-sectional area and length, and usually at a given temperature. In practical electrical work, the resistivity is specified as a *circular mil foot* at 20°C. For example, as shown in Fig. 2–1, the resistivity of copper is 10.4 at 20°C. This means that 1 ft of copper wire 1 mil in diameter produces 10.4 Ω of resistance at a temperature of 20°C (68°F).

2-1.2 Circular Mils versus Square Mils

The circular mil (or *cmil*) is chosen for electrical work since most conductors or wires are round (and small). A mil is one thousandth of an inch (0.001 in.). A circular mil is a circle 0.001 in. in diameter.

As shown in Fig. 2–1, to find the area of a round conductor in circular mils, simply square the diameter. For example, if the diameter of a wire is 0.050 in., or 50 mils, the area in cmils is 2500 (50 × 50).

Most of the tables and charts found in this book (and those used in the NEC and most electrical equipment manufacturers) use circular mils to describe wire sizes. However, there are cases where you must convert cmils to a square measure, or vice versa. This can be done simply by the use of the following equations:

$$\text{cmils} = \text{square mils} \times 1.273 \quad \text{square mils} = \text{cmils} \times 0.7854$$

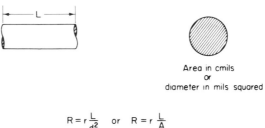

$$R = r \frac{L}{d^2} \quad \text{or} \quad R = r \frac{L}{A}$$

R = Resistance of wire
r = Specific resistance of material
L = Length of wire in feet
d = Diameter in mils
A = Area (in cmils)

Material	Resistivity at 20°C (68 °F) (Ω–cmil/foot)
Silver	9.8
Copper	10.4
Aluminum	17
Tungsten	33
Nickel	50
Iron	60
Manganin	290
Nichrome	660

FIGURE 2-1   Resistance of wire.

For example, assume that a bus bar is $\frac{1}{4}$ in. by 2 in. The square-mil area is found as follows:

$$\frac{1}{4} \text{in.} = 250 \text{ mils}$$
$$2 \text{ in.} = 2000 \text{ mils}$$
$$250 \times 2000 = 500,000$$
$$\text{cmils} = 500,000 \times 1.273 = 636,500 \text{ square mils}$$

2-1.3 Calculating Resistance of Conductors

In designing electrical wiring, it is often necessary to find the resistance of conductors or wires. For example, many design problems require that the voltage drop produced by a given length be known. If the resistance for that length of wiring is known, the voltage drop can be calculated for a given current flow or power consumption.

As shown in Fig. 2-1, the resistance of a conductor increases as the length increases, and decreases as the cross-sectional area increases (the wire size increases). For example, find the resistance of 1000 ft of copper wire with a diameter of 0.050 in., at a temperature of 20°C, using the equations and data of Fig. 2-1. (Note that 0.050 in. is the approximate diameter of No. 16 AWG wire.) D = 0.050 in. or 50 mils, L = 1000 ft, and resistivity = 10.4, so

$$\frac{10.4 \times 1000}{50^2} = \frac{10,400}{2500} = 4.16 \ \Omega$$

2-1.4 Conductor (Wire) Size

The sizes of wires (or conductors) used in practical electrical work conform to the American Wire Gauge (AWG) system. Under the AWG system, the smallest wire is No. 40, which has a diameter of 3.145 mils (much too small for electrical wiring, but used in electronics). The AWG numbers then decrease (one number at a time) as the wire size increases, up to No. 0000 AWG (also described as 4/0, or 4-zero), which is almost $\frac{1}{2}$ in. in diameter (460 mils to be exact). Wires larger than No. 0000 AWG are described in circular-mil area. To simplify the list of the larger wire sizes, the wire is described in *thousands of circular mils* (which is abbreviated as either *kcmil,* or as MCM, depending on the literature). The next size larger than No. 0000 AWG is 250 kcmil or MCM 250. Either way, this means 250,000 circular mils, or 500 mils (exactly $\frac{1}{2}$ in.) in diameter.

The wire-size system is shown in Fig. 2-2, which lists data for copper and aluminum wires in sizes from No. 18 AWG to 1000 kcmil or, MCM 1000 (1 in. in diameter). These are the *standard sizes* used in most practical electrical wiring systems. Note that Fig. 2-2 gives the area in cmils, and the diameter of each wire in mils. The diameter can be converted to inches by moving the decimal point three places to the left. For example, the diameter of No. 16

Size	Area (cmil)	Number of wires	Diameter each wire (in.)	D-c resistance at 25°C (77°F) (Ω/1000 ft)	
				Copper	Aluminum
AWG					
18	1,620	Solid	0.0403	6.51	10.7
16	2,580	Solid	0.0508	4.10	6.72
14	4,110	Solid	0.0641	2.57	4.22
12	6,530	Solid	0.0808	1.62	2.66
10	10,380	Solid	0.1019	1.018	1.67
8	16,510	Solid	0.1285	0.6404	1.05
6	26,240	7	0.0612	0.410	0.674
4	41,740	7	0.0772	0.259	0.424
2	66,360	7	0.0974	0.162	0.266
1	83,690	19	0.0664	0.129	0.211
0	105,600	19	0.0745	0.102	0.168
00	133,100	19	0.0837	0.0811	0.133
000	167,800	19	0.0940	0.0642	0.105
0000	211,600	19	0.1055	0.0509	0.0836
kcmil or MCM					
250	250,000	37	0.0822	0.0431	0.0708
300	300,000	37	0.0900	0.0360	0.0590
350	350,000	37	0.0973	0.0308	0.0505
400	400,000	37	0.1040	0.0270	0.0442
500	500,000	37	0.1162	0.0216	0.0354
600	600,000	61	0.0992	0.0180	0.0295
700	700,000	61	0.1071	0.0154	0.0253
750	750,000	61	0.1109	0.0144	0.0236
800	800,000	61	0.1145	0.0135	0.0221
900	900,000	61	0.1215	0.0120	0.0197
1000	1,000,000	61	0.1280	0.0108	0.0177

FIGURE 2-2   Conductor resistance in ohms per 1000 ft.

AWG is 50.8 mils, which is 0.0508 in. Also note that wire sizes No. 8 through No. 18 AWG are generally solid wires, whereas No. 6 AWG and larger are *stranded* wires. The diameter column of Fig. 2–2 lists the diameter of *each strand* in stranded wires. This diameter must be squared, and then multiplied by the number of strands, to find the area in cmils. Of course, the area column lists the cmil area for each wire size, so this need not be calculated. The diameter and number of wire strands are given only for reference.

Figure 2–2 lists the resistance of both copper and aluminum conductors in each wire size. A glance at the resistance columns brings up several points to be remembered in practical electrical wiring. Obviously, the larger the wire, the less resistance—all other factors (length, temperature, material) being equal. Next, aluminum has more resistance than copper. *As a guideline, use the next larger standard wire size when aluminum is used instead of copper.* For example, the resistance for 1000 ft of aluminum No. 16 AWG is approximately the same as for 1000 ft of copper No. 18 AWG. This is a guideline only, and applies primarily to the smaller wire sizes found in small-building wiring.

The resistance columns give the resistance for 1000 ft of conductor at a temperature of 25°C (77°F). This is different from the 20°C temperature of Fig. 2–1. If you bother to calculate the resistance using both Figs. 2–1 and 2–2, you will note a slight difference in resistance (resulting from the small difference in the two temperatures). However, if there is a large difference in temperature, there is a corresponding difference in resistance (all other factors remaining the same).

2-1.5 Effects of Temperature on Conductor Resistance

The resistance of any conductor varies with temperature. For conductors used in practical electrical wiring, the resistance increases when temperature increases. Typically, the resistance of conductors is listed at temperatures in the range 20 to 25°C (68 to 77°F). This range is considered the normal or average operating temperature of electrical wiring. The actual operating temperature of wiring is controlled by two major factors. First, there is the ambient temperature, or temperature that surrounds the wiring. Second, there is the temperature that results from heat caused by current flowing in the wires. As current increases, the generated heat increases, as does the resistance. The increase in resistance tends to decrease current. (This is fortunate since, if the opposite were true, the wiring would probably burn out!)

In theory, if the temperature of any conductor is lowered far enough, the resistance is zero. In practical work, this is impossible. For one reason, if there is any current flow, there is some heat, and an increase in temperature (thus producing an increase in resistance). However, there is an *inferred zero resistance temperature* for conductors, shown in Fig. 2–3 for several common conductors together with equations for finding resistance at different temper-

$$R_1 = R_2 \frac{IZR + T_1}{IZR + T_2} \qquad R_2 = R_1 \frac{IZR + T_2}{IZR + T_1}$$

R_1 = resistance at low temperature

R_2 = resistance at high temperature

T_1 = low temperature

T_2 = high temperature

IZR = inferred zero resistance temperature

Inferred temperature for zero resistance (C°)	Material
−243	Silver
−234.5	Copper
−236	Aluminum
−202	Tungsten
−147	Nickel
−180	Iron
−6250	Nichrome

FIGURE 2-3 Determining conductor resistance at different temperatures.

atures. To use Fig. 2–3, it is necessary to know the resistance of the conductor at a particular temperature. Armed with this information, you can find the resistance of the conductor at any other temperature.

For example, Fig. 2–2 shows a resistance of 4.1 Ω for 1000 ft of No. 16 AWG copper wire at 25°C. Assume that the temperature increases to 50°C. Find the resistance for the same wire using the equations of Fig. 2–3.

$$\text{resistance at } 50°C = 4.1 \times \frac{234.4 + 50}{234.5 + 25} = 4.5 \ \Omega \ \text{(approx.)}$$

Now assume that the temperature drops to 0°C. Find the resistance for the same wire.

$$\text{resistance at } 0°C = 4.1 \times \frac{234.5 + 0}{234.5 + 25} = 3.8 \ \Omega \ \text{(approx.)}$$

2-1.6   Skin Effect

Thus far, we have discussed the resistance of conductors in reference to direct current (dc). In practical electrical wiring problems, we must deal with alternating current (ac). This brings up the problem of *skin effect*. When alternating current flows in a conductor, the current tends to flow on the outside, or "skin," of the conductor. This decreases the conductor area and increases resistance. That is, the a-c resistance of a conductor is higher than the d-c resistance, all other factors being equal. As the frequency of the ac increases, the skin effect becomes more pronounced and the resistance increases.

From a practical standpoint, skin effect is a problem only on larger conductors (typically on conductors of No. 1 AWG or larger). Also, skin effect is more pronounced when conductors are enclosed in metal (such as metal-clad cable, rigid conduit, etc.).

Figure 2–4 shows multiplying factors for converting d-c resistance to a-c resistance (at the commonly used frequency of 60 Hz) for both copper and aluminum. Note that skin effect is more pronounced for copper than aluminum.

To use Fig. 2–4, multiply the d-c resistance by the multiplying factor. For example, Fig. 2–2 shows that the resistance for 1000 ft of MCM 800 copper conductor is 0.0135 Ω. Assume that this same 1000 ft of copper conductor is used with 60-Hz ac, enclosed in a metal raceway. Find the a-c resistance.

$$\text{dc resistance} \times \text{multiplier} = \text{ac resistance}$$
$$0.0135 \times 1.22 = 0.01647 \ \Omega$$

Size AWG or MCM	Conductors in metal cables or raceways		Conductors in air or non-metal cables and raceways	
	Copper	Aluminum	Copper	Aluminum
AWG 3	1	1	1	1
2	1.01	1	1	1
1	1.01	1	1	1
0	1.02	1	1.001	1
00	1.03	1	1.001	1.001
000	1.04	1.01	1.002	1.001
0000	1.05	1.01	1.004	1.002
MCM 250	1.06	1.02	1.005	1.002
300	1.07	1.02	1.006	1.003
350	1.08	1.03	1.009	1.004
400	1.10	1.04	1.011	1.005
500	1.13	1.06	1.018	1.007
600	1.16	1.08	1.025	1.010
700	1.19	1.11	1.034	1.013
750	1.21	1.12	1.039	1.017
800	1.22	1.14	1.044	1.017
900	1.27	1.17	1.056	1.022
1000	1.30	1.19	1.067	1.026

DC Resistance • Multiplier = AC Resistance

FIGURE 2-4   Skin-effect multiplying factor.

2-2   CONDUCTOR MATERIALS AND INSULATION

Copper or aluminum are used as conductors for almost all modern electrical wiring. Of the two, copper is most frequently used for all sizes of wire and is used almost exclusively for the small wire sizes. Each conductor material has its advantages.

Copper has less resistance (10.4 Ω/cmil ft, compared to 17 Ω/cmil ft for aluminum) and is stronger. Copper thus carries more current for a given wire size.

Aluminum is less expensive and is lighter (about one-third lighter than copper). Thus a larger aluminum conductor can weigh less (and cost less) than a smaller copper conductor of the same current capacity.

No matter what material is used, all conductors must have some form of insulation. The type of insulating material is set by the conditions or environment in which the conductor is used. Heat and moisture are the main problems. Excessive heat (caused either by external conditions or high currents, or both) can melt insulation; extreme heat can burn insulation. Excessive moisture can penetrate some insulation and result in a short circuit. Also, age can cause insulation to deteriorate. For example, age and heat cause natural rubber to crack. The amount of insulation is set by the voltage between conductors. A higher voltage requires more insulation.

The NEC has classifications for types of insulating materials and for voltages. There are six general voltage classifications for insulation: 600, 1000, 2000, 3000, 4000, and 5000 V. The NEC requires that conductors be identified by the first two numbers stamped or printed on the insulation. For example, the numbers 10 indicate that the insulation is suitable for voltages up to 1000

Type	Material and characteristics	Application	Maximum operating temperature (°F)
R	Rubber	Dry locations	140
RH	Heat-resistant rubber	Dry locations	167
RHH	Higher-temperature, heat-resistant rubber	Dry locations	194
RHW	Heat- and moisture-resistant rubber	Dry or wet locations	167
T	Thermoplastic	Dry locations	140
TH	Heat-resistant thermoplastic	Dry locations	167
THW	Heat- and moisture-resistant thermoplastic	Dry or wet locations	167
THWN[a]	Heat- and moisture-resistant thermoplastic with nylon covering	Dry and wet locations	167

[a]THWN is thinner than other insulations in general use, and the overall size of THWN conductor (including insulation) is smaller than conductors with other insulating materials of the same size number.

FIGURE 2-5 Basic classifications of insulation types in general use.

V maximum. When there is no identifying number, the conductor insulation is for use with 600 V maximum.

There are several classifications of insulation types in general use. Each type is assigned an identifying letter. The letters indicate insulation material or application, or both. There are five letters for the type classifications: R for rubber, T for thermoplastic, N for nylon, H for heat resistant, and W for moisture resistant. The basic NEC classifications are shown in Fig. 2-5.

2-3 CONDUCTOR CURRENT CAPACITY (AMPACITY)

The word *ampacity* means capacity in amperes. When applied to conductors, the term "ampacity" means the maximum current, in amperes, that can be carried in a conductor *with safety*. The NEC provides tables that specify the current-carrying capacity, or ampacity, of commonly used conductor sizes, under a given set of conditions. Figures 2-6 and 2-7 are examples of such tables. The information shown in Figs. 2-6 and 2-7 should be used for convenience. However, always check any final calculations against NEC tables (of the latest issue).

Figure 2-6 shows the ampacity of single insulated conductors in free air. Note that the ampacity is set by the type of insulation and the conductor material, as well as by the conductor size. For example, for No. 6 AWG copper conductors, the ampacity is 60 A, when R, T, or TW insulations are used, but is increased to 95 A for RH, RHW, TH, and THW insulations. Using the

Size AWG or MCM	Copper R,T,TW, 60°C(140°F)	RH,RHW,TH THWN,THW 75°C(167°F)	Aluminum R,T,TW 60°C(140°F)	RHW,TH THW,THWN 75°C(167°F)
AWG				
14	20	20	—	—
12	25	25	20	20
10	40	40	30	30
8	55	65	45	55
6	80	95	60	75
4	105	125	80	100
3	120	145	95	115
2	140	170	110	135
1	165	195	130	155
0	195	230	150	180
00	225	265	175	210
000	260	310	200	240
0000	300	360	230	280
MCM				
250	340	405	265	315
300	375	445	290	350
350	420	505	330	395
400	455	545	355	425
500	515	620	405	485
600	575	690	455	545
700	630	755	500	595
750	655	785	515	620
800	680	815	535	645
900	730	870	580	700
1000	780	935	625	750

FIGURE 2-6 Ampacity of single insulated conductors in free air.

Size AWG or MCM	Copper R,T,TW 60°C(140°F)	RH,RHW,TH THW,THWN 75°C(167°F)	Aluminum R,T,TW 60°C(140°F)	RH,RHW,TH, THW,THWN 75°C(167°F)
AWG				
14	15	15	—	—
12	20	20	15	15
10	30	30	25	25
8	40	45	30	40
6	55	65	40	50
4	70	85	55	65
3	80	100	65	75
2	95	115	75	90
1	110	130	85	100
0	125	150	100	120
00	145	175	115	135
000	165	200	130	155
0000	195	230	155	180
MCM				
250	215	255	170	205
300	240	285	190	230
350	260	310	210	250
400	280	335	225	270
500	320	380	260	310
600	355	420	285	340
700	385	460	310	375
750	400	475	320	385
800	410	490	330	395
900	435	520	355	425
1000	455	545	375	445

FIGURE 2-7 Ampacity of three (or less) conductors in a cable or raceway.

same No. 6 AWG size and an aluminum conductor, the ampacity ratings are reduced to 60 and 75 A.

The information of Fig. 2-6 is not too realistic, in that most electrical wiring does not involve a single conductor in free air. Typically, three (or at least two) conductors are used in cables or raceways.

Figure 2-7 shows the ampacity of three (or less) conductors in a cable or raceway. Again, the ampacity is set by insulation, conductor material, and size. However, the ampacity of conductors in cables or raceways is always less than for a single conductor in free air. For example, using the same No. 6 AWG size, the ampacity for copper conductors is reduced to 55 and 65 A.

2-3.1 Temperature Derating Factors

Although the information in Fig. 2-7 can be used for practical work, it has certain limitations. The figures in Fig. 2-7 are based on operating the electrical wiring in an ambient temperature not over 30°C (86°F). If the normal ambient temperature is higher than 30°C, the conductor ampacity must be *derated*. That is, a lower current must be used for reasons of safety.

Figure 2-8 shows derating factors for ampacity of conductors used in ambient temperatures over 30°C. Note that the derating factor for conductors with insulations normally used in lower temperatures is different than that for high-temperature insulations. For example, assuming an ambient temperature of 40°C (104°F), the derating factor is 0.82 for R, T, and TW insulations, and is 0.88 for RH, RHW, TH, THW, and THWN insulations.

To find the ampacity of a conductor operating at an elevated temperature (above 30°C), *multiply* the 30° ampacity (Fig. 2-7) by the derating factor.

Conductor derating factor				
R, T, TW		RH, RHW, TH, THW, THWN		Elevated ambient temperature [°C (°F)]
Factor	Reciprocal	Factor	Reciprocal	
0.82	(1.22)	0.88	(1.14)	40 (104)
0.71	(1.41)	0.82	(1.22)	45 (113)
0.58	(1.73)	0.75	(1.33)	50 (122)
0.41	(2.44)	0.67	(1.5)	55 (131)
—	—	0.58	(1.73)	60 (140)
—	—	0.35	(2.86)	70 (158)
—	—	0.20	(5.0)	75 (167)

FIGURE 2-8 Conductor derating factors (and reciprocals) for elevated temperatures.

For example, as shown in Fig. 2–7, the ampacity of No. 8 AWG copper conductor with R, T, or TW insulation is 40 A. If this same conductor is operated at 55°C (131°F), the ampacity is reduced to 16.4 A (40 × 0.41), as shown in Fig. 2–8.

Note that the numbers of Fig. 2–8 in parentheses are reciprocals of the derating factor. These reciprocals are convenient since the problem of ampacity derating factors is generally stated in reverse, in practical electrical work. That is, a given load requires a given current and, if the wiring is normally operated at temperatures above 30°C, a large wire size or a different insulation is required. That means that you must divide the required current by the derating factor to find a "derated current," and then use the higher derated current to find the correct conductor size and insulation type.

For example, assume that the load requires 100 A and that three TW copper conductors are to be operated in a cable at ambient temperatures of 55°C.

The first step is to divide the 100 A by the TW derating factor for 55°C, which is 0.41 (Fig. 2–8). Since it is easier to multiply than divide, use the reciprocal (2.44 in this case) to find a derated current of 244A (100 A × 2.44).

Next, check the TW column for copper conductors in Fig. 2–7 to find the conductor size that accommodates 244 A. Always use the *next larger size,* unless an exact size can be found. In this case, 350 MCM accommodates 260 A (for copper with TW insulation). A size of 300 MCM is slightly low (for 240 A), so the 350 MCM should be used.

As an alternative, practical solution, if the insulation can be changed to THW, the size can be reduced to No. 0 AWG. This drastic reduction in size is possible because of the different derating factor of THW. As shown in Fig. 2–8, the derating factor for THW insulation at 55°C is 0.67, and the reciprocal is 1.5. Using the reciprocal, multiply the required current of 100 A by 1.5 to find a derated current of 150 A. Then check the THW column for copper conductors in Fig. 2–7 to find the conductor size that accommodates 150 A. In this case, No. 0 AWG does the job.

2–3.2 Derating Factors for Multiple Conductors

The information in Fig. 2–7 is based on using a maximum of three conductors in a raceway or cable. This applies to many commercial and industrial installations (small offices, stores, etc.). However, in larger electrical systems, it is generally more convenient (and economical) to have many conductors in a raceway. When more than three conductors are used in the same raceway or cable, the conductor ampacity must be derated, so that a large conductor is required for safety.

Figure 2–9 shows derating factors for ampacity of multiple conductors (more than three) used in the same raceway or cable. To find the ampacity of a *single conductor* operating in a raceway or cable with more than three con-

Conductor derating factor		
Factor	Reciprocal	Number of conductors in raceway
0.8	1.25	4 to 6
0.7	1.43	7 to 24
0.6	1.67	25 to 42
0.5	2.0	43 or more

FIGURE 2-9   Conductor derating factors for more than three conductors in a raceway.

ductors, multiply the single-conductor ampacity (Fig 2–6) by the derating factor. For example, as shown in Fig. 2–6, the ampacity of No. 8 AWG copper conductor with R, T, or TW insulation is 55 A. If this same conductor is operated in a raceway with 7 to 24 conductors, the ampacity is reduced to 38.5 A (55 × 0.7).

To find the ampacity of a three-wire system operating in a raceway or cable with more than three conductors, *multiply* the three-conductor ampacity (Fig. 2–7) by the derating factor. For example, as shown in Fig. 2–7, the ampacity of No. 8 AWG copper conductor with R, T, or TW insulation in a three-conductor system is 40 A. If these same conductors are operated in a raceway with 7 to 24 conductors, the ampacity is reduced to 28 A (40 × 0.7).

Again, for practical work, the problem of multiple-conductor derating is usually stated in reverse. A given load requires a given current, and a larger wire size is required when multiple conductors are used.

For example, assume that the load requires 100 A and that three TW copper conductors are to be operated in a cable with a total of 7 to 24 conductors.

The first step is to divide the 100 A by the 7 to 24 conductor derating factor (0.7), or multiply by the reciprocal (1.43). Using the reciprocal, multiply the required current of 100 A by 1.43 to find a derated current of 143 A. Then check the TW column for copper conductors in Fig. 2–7 to find the conductor size that accommodates 143 A. In this case, No. 00 AWG will do the job.

2-3.3 Three-Phase Derating Factors

Several special derating factors must be considered when three-phase wiring is used.

In three-phase four-wire systems that are balanced, the fourth (or neutral) wire carries no current and produces no heat. Thus balanced three-phase four-wire systems can be rated the same as three-wire systems, as far as ampacity of conductors is concerned. Even if the three-phase four-wire systems are unbalanced, the line currents decrease as neutral currents increase. Therefore, the net heating effect is the same, and the information of Fig. 2–7 applies.

If a three-phase four-wire system is used with fluorescent lighting (Chapter 7), the neutral conductor current is not zero. This applies even if the three-phase four-wire system is balanced. Thus in any three-phase four-wire system involving fluorescent lights, the wire sizes must be derated in ampacity using the factors of Fig. 2-9.

2-3.4 Derating Factors for Neutral Wires

The NEC allows the size of the neutral wire to be reduced if the current of the neutral wire in a three-phase system (or single-phase system) is over 200 A.

The size of the neutral wire is based on a current equal to 200 A, plus 70% of the current over 200 A. For example, if the maximum current in the neutral wire is 300 A, the current used to find the correct wire size is 270 A (300 A − 200 A = 100 A; 100 A × 70% = 70 A; 200 A + 70 A = 270 A). However, *the NEC does not permit this reduction if any of the load is comprised of fluorescent lighting.*

2-4 CONDUCTOR VOLTAGE DROP

Since all conductors have some resistance, there is always some voltage drop (or difference) between the voltage source and the load. Since there is both voltage drop and resistance, there is some power consumption in any conductor. As a result, all conductors generate some heat.

From a practical wiring standpoint, the main concern is that the voltage drop does not exceed a certain percentage of the source voltage. For example, *the NEC permits a maximum of 3% voltage drop to the farthest outlet in a branch circuit, and a 5% voltage drop if feeders are included.* (Feeders and branch circuits are discussed in Chapter 4.)

In a typical two-wire 120-V system, the maximum voltage drop that can be tolerated from the service entrance to the farthest outlet is 6 V (120 × 5%). Keep in mind that this represents a voltage drop of 3 V (or one-half of 6 V) for each of the two conductors, as shown in Fig. 2-10.

For a given wiring design problem, a larger wire size produces a larger cmil area and a lower resistance (a larger wire size produces a lower voltage drop). A simple solution to any voltage drop problem is to use the largest

FIGURE 2-10 Maximum conductor voltage drop in 120-V two-wire system.

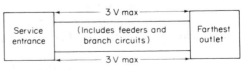

possible wire size, but this is neither practical nor economical. As any contractor knows, an increase in one or two wire sizes can sometimes raise the overall cost of a job to a point where it is no longer competitive.

A more practical solution is to use a wire size that produces a voltage drop *just below* the maximum permitted. Of course, the wire size selected for voltage drop must also be capable of carrying the maximum rated current.

2-4.1   *Finding Correct Wire Size to Provide a Given Voltage Drop*

Although there are several methods for finding a wire size that produces a given voltage drop, the following is most direct. This method involves finding the area *in cmils* that produces the voltage drop, and choosing the *next largest standard wire size*. Then the selected wire size is checked for ampacity, and derated (if necessary) for heat, number of conductors, skin effect, and so on, as described in Secs. 2–1 through 2–3.

The basic equation for finding the cmil area for a given voltage drop is

$$\text{cmil} = \frac{\text{resistivity of material} \times \text{current} \times \text{distance}}{\text{source voltage} \times \text{percentage of drop}}$$

For single-phase, the distance is two times the distance from source to load. This applies to both two- and three-wire single-phase systems, since the neutral wire carries no current for balanced loads. If the load is unbalanced, the neutral wire carries less current than either of the other wires. Thus the wire size selected for the other wires produces a lower voltage drop in the neutral wire.

For three-phase, the distance is 1.732 times the distance from source to load.

The load is usually expressed in terms of power (watts or kilowatts), rather than in current (amperes), for most practical problems. For single-phase, convert power to current using the equation $I = P/E$, using the *maximum load* in watts for P and the source voltage for E. For three-phase, the conversion is more complex: $I = P/E \times 1.732 \times \text{power factor}$. (Refer to Chapter 9.)

Example of Conductor Voltage Drop in a Single-Phase System.   Assume that copper conductors are required to supply a 100-A load 300 ft from a 120-V single-phase source with a voltage drop not to exceed 3%.

You start by finding the correct cmil area of the conductors (which, being copper, have a resistivity of 10.4, as shown in Fig. 2–1).

$$\text{cmil} = \frac{10.4 \times 100 \times (300 \times 2)}{120 \times 0.03} = 173{,}333$$

Next, find the correct wire size (nearest larger size), which happens to be No. 0000 AWG, as shown in Fig. 2–2.

Now check the standard wire size you have selected for correct ampacity. As shown in Fig. 2–7, No. 0000 copper conductors with any type of insulation can carry more than 100 A. Of course, if the conductors are to be operated at high temperatures, or with many conductors in a raceway, the derating factors (Sec. 2–3) must be applied. Since conductors of a larger size are involved, it is also wise to check the skin effect (Sec. 2–1.6).

The maximum permitted voltage drop is 3.6 V (120 × 0.03) with 100 A of current. This results in a maximum d-c resistance of 0.036 Ω ($R = E/I$; 3.6/100 = 0.036). The total length of the conductor is 600 ft (300 ft from source to load, and 300 ft from load back to source). Thus the maximum allowable resistance for 600 ft of conductor is 0.036 Ω. Using the information of Sec. 2–1.3, the d-c resistance for 600 ft of No. 0000 AWG conductor with an assumed resistivity of 10.4 is

$$\frac{10.4 \times 600}{211,600} = 0.0295 \ \Omega$$

A quicker, but slightly less accurate method for finding the resistance of the conductor is to use the d-c resistance columns of Fig. 2–2 for 1000 ft, and then correct for the actual number of feet. For example, Fig. 2–2 shows that 1000 ft of No. 0000 AWG copper conductor has a resistance of 0.0509 Ω. Multiply this 0.0509 by 0.6 (for 600 ft) to obtain a resistance of about 0.03 Ω.

No matter which calculation you use, the 0.03 or 0.0295 Ω is less than the maximum permitted d-c resistance of 0.036 Ω. However, as shown in Fig. 2–4, a multiplying factor of 1.05 must be applied to copper No. 0000 AWG conductors in a metal raceway, for skin effect. This makes the a-c resistance 0.0315 Ω (0.03 × 1.05), which is still below the maximum permitted 0.036 Ω.

Example of Conductor Voltage Drop in a Three-Phase System. Now assume that we have the same load (12 kW; 120 V × 100 A), and the same distance from voltage source to load (300 ft). However, we decide to use aluminum conductors, and 240-V three-phase, with a power factor of 0.8. The voltage drop must still not exceed 3.6 V (which is 1.5% of 240 V).

First convert the 12-kW load to a current:

$$I = \frac{12,000}{240 \times 1.732 \times 0.8} = \text{approximately 36 A}$$

With a current of 36 A, find the cmil area:

$$\text{cmil} = \frac{17 \times 36 \times (300 \times 1.732)}{240 \times 0.015} = \text{approximately 88,332}$$

The next larger standard size is No. 0 AWG (Fig. 2–2). Check this standard size for correct ampacity. As shown in Fig. 2–7, No. 0 AWG aluminum conductors with any type of insulation can carry more than 36 A. Check the derating factors as described in Secs. 2–1 through 2–3, paying particular attention to Sec. 2–3.3. Skin effect can be ignored, because the skin-effect multiplying factor (Fig. 2–4) for No. 0 AWG aluminum conductor is 1.

2–5   THE NEED FOR RACEWAYS

Electrical conductors must be protected from heat, moisture, and physical abuse. Of course, part of this protection is provided by insulation, but that is not enough. For example, should insulation become even slightly frayed (because of heat or abuse) there is an immediate shock hazard. If there is moisture near the frayed insulation, the moisture can conduct electrical current and cause the wiring to malfunction (or possibly burn).

These possibilities are minimized by installing all conductors in race-

FIGURE 2–11   Typical wireway used for industrial applications. (Courtesy Square D Company.)

ways, which can be considered as enclosed channels for wiring. Because of their importance, *the NEC requires that all conductors be enclosed and/or protected by some form of raceway.*

There are three basic types of raceways used in commercial and industrial work. The most common raceway is the familiar *conduit* or *cable* (Sec. 2–6), which is used when a limited number of wires (and low ampacities) are involved. Some form of *wireway* (Sec. 2–7) is used to house the conductors when the number of wires and ampacity must be increased. Figure 2–11 shows a typical wireway installation, combined with rigid conduit. When commercial or industrial installations involve very heavy currents (exceeding those values given in Figs. 2–6 and 2–7), *bus bars* can be used instead of wires. In most modern installations, bus bars are housed in *busways* (Sec. 2–8). Such busways are essentially metal ducts with built-in bus bars, and openings at various points along the duct. The openings permit switches and outlets to be plugged into the bus bars as required. One of the main advantages of busways is that they permit the addition of outlets without extensive rewiring.

The choice of raceway type is sometimes set strictly by local electrical codes such that, in effect, there is no choice. In other cases, there are several approved raceway types, so that cost is the determining factor. (Note that this cost factor involves both the cost of the raceway material and the cost of labor for installing the raceway.) The next three sections summarize the types of raceways available for commercial and industrial installations.

2-6 CONDUIT, TUBING, AND CABLE

Metal tubing of one form or another has long been used for raceways. There are two types of tubing in common use as raceways: rigid conduit and electrical metallic tubing.

2-6.1 *Rigid Conduit*

Rigid conduit looks like plumber's pipe and must be handled in about the same way. That is, rigid conduit must be bent to the desired routing (using plumber's pipe-bending tools) and must be threaded at each end. Rigid conduit is attached to other raceway equipment (switch boxes, outlets, ducts, etc.) using bushings and locknuts on the threaded ends, as shown in Fig. 2–11. Generally, rigid conduit is available in 10-ft lengths. The main difference between rigid conduit and plumber's pipe is that the interior of the conduit has a heavy enameled finish. This smooth finish permits the conductors to be pulled through the conduit with a minimum of effort.

The main advantage of rigid conduit is strength. Like plumber's pipe, rigid conduit can be used in areas where there is a chance of physical abuse or where there is extreme moisture. Unlike many other types of raceways, rigid conduit can be buried in concrete when required.

2-6.2   Electrical Metallic Tubing

Electrical metallic tubing (EMT) is similar to rigid conduit except that EMT is much thinner and cannot be threaded at the ends. EMT (sometimes known as *thin wall*) is secured to other raceway components by compression rings and/or setscrews. *The maximum permitted diameter of* EMT is 4 in. EMT is almost as strong as rigid conduct, and EMT can be buried in concrete. However, EMT *is not to be used where there is continuous moisture.* Special bending tools are required for EMT.

2-6.3   Metal-Clad (Armored) Cable

There are two types of metal-clad or armored cables (Fig. 2–12) used as raceways: AC and MC. Both types are manufactured as complete cables with the conductors already installed. This eliminates the need to pull the conductors through the raceway. The outer shield of the cables is formed of a spiral steel wrapping. This type of construction makes the raceway, and the conductors, into a flexible cable that can be secured by clamps at various points along its route. The metal-clad cables are attached to switch boxes, outlets, and so on, by means of clamps. Special care must be taken to protect the conductors from the sharp edges of the spiral wrapping at the point where the metal-clad cable terminates. Special bushings and fittings are usually required for this purpose.

Metal-clad cable cannot be buried in concrete. One of the reasons for raceways is that conductors can be replaced easily. In metal-clad cable, the conductors and raceway are an inseparable unit, and it is impossible to replace the conductors by pulling them through the armor wrapping. Regular metal-

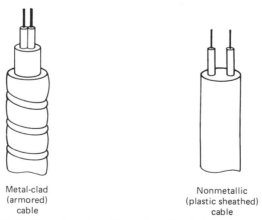

Metal-clad
(armored)
cable

Nonmetallic
(plastic sheathed)
cable

FIGURE 2-12   Construction of typical electrical cables.

clad cable is not to be used in areas of continuous moisture. However, there is a special metal-clad cable with lead-covered conductors that can be used in high-moisture areas. This metal-clad cable is designated ACL.

The designations AC and MC in metal-clad cable refer to size. MC is used for large loads only, since the smallest conductors available in MC cables are No. 4 AWG for copper and No. 2 AWG for aluminum. AC is used for smaller loads and is available with conductors as small as No. 14 AGW. *Note that AC cable is often referred to as BX cable.*

Typically, BX cable is available in two-conductor and three-conductor assemblies. With either type, one conductor must be white (to identify it as the grounded conductor, as discussed in Chapter 5). Generally, the other conductor in a two-conductor cable is black or red. In three-conductor cables, the other conductors are usually black and red or orange and blue.

2-6.4 Nonmetallic Cable (Romex)

Nonmetallic cable (Fig. 2–12) is generally called *Romex,* although not all such cable may be true Romex. Nonmetallic cable is also manufactured as a complete assembly with the conductors already installed. However, the outer shield is made of plastic or a heavy fabric. Nonmetallic cable is installed in the same way as metal-clad cable but will not take physical abuse without damage. *For this reason, nonmetallic cable does not meet the requirements of many local codes.*

For example, nonmetallic cable cannot be used where there is any possibility of physical abuse or in high-moisture areas. Since the outer shield or armor is nonmetallic, the shield cannot conduct current. Thus the shield cannot be used to conduct ground-fault currents, as described in Chapter 5. To overcome this problem, some nonmetallic cable is provided with an extra conductor (usually smaller than the remaining conductors) for grounding purposes.

Typically, nonmetallic cable is available in two-conductor and three-conductor assemblies, as is metal-clad cable. Nonmetallic cable is generally available with conductors sizes No. 14 to 1 AWG (copper) and No. 12 to 2 AWG (aluminum).

2-7 WIREWAYS

A typical wireway is a steel-enclosed trough or duct designed to carry electrical wires and cables, and to protect the conductors against possible damage. In effect, wireways are special-purpose raceways designed to meet a number of specific applications. For example, there are wireways called *surface extensions,* decorative sheet-metal channels designed for routing wires along surfaces (such as walls, floors, overheads, etc.). Generally, surface extensions are used as extensions of other raceways rather than as full raceway systems designed to wire a complete building.

We make no effort to describe all types of wireways here. Instead, we concentrate on wireways manufactured by Square D Company. The Square D line includes general-purpose, raintight, underfloor duct, and industrial wireways [manufactured to JIC (Joint Industry Conference) specifications]. Figure 2–13 shows a typical industrial wireway installation. The remainder of this section describes the main features of the Square D wireway line, together with the design capabilities and limitations.

2-7.1 *SQUARE-Duct*®

SQUARE-Duct®, shown in Fig. 2–14, is a steel-enclosed wireway with a removable hinged cover that can be used as either hinge-cover wireway or screw-cover trough. A complete set of fittings is available so that an entire wireway system can be installed, regardless of bends, offsets, or other building contours that may be encountered. Fittings have removable covers and sides to permit a complete "lay-in" installation, and to provide access to wires through the entire length, without any alterations to the system. Lengths can be opened, wires laid in, and wireway closed quickly. The cover can be removed to make connections easier, in tight spots. SQUARE-Duct is UL listed as wireways and auxiliary gutters.

SQUARE-Duct wireway can be used to advantage in most smaller distribution systems where multiple runs of conduit are required, and is superior to conduit for exposed work, or where additions or alterations to the distri-

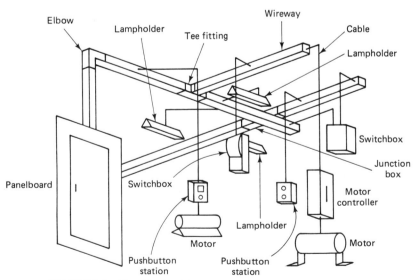

FIGURE 2-13   Typical industrial wireway installation.

FIGURE 2-14 SQUARE-Duct wireway. (Courtesy Square D Company.)

bution system can be expected. Installation can be made without expensive tools. Normally, a screwdriver and wrench are all that are needed to install both lengths and fittings. Runs are assembled on the floor and hand-lifted into position.

Wireway systems installed in SQUARE-Duct are adaptable and readily accessible at all times. Taps and splices can be made whenever an apparatus is to be added, moved, or modified. It is simple to add or reroute a circuit anytime after the original installation is completed. SQUARE-Duct systems are completely reusable and there is no need to scrap a major portion of the system when it is necessary to revise the layout.

2-7.2 Fittings

Figures 2-15 through 2-18 show some typical fittings for SQUARE-Duct wireways. The connectors (Fig. 2-15) form a closed section with a friction hinge on one side. This hinge prevents the connector cover from falling closed, and holds the cover open for easy lay-in of wire and cable where the cover cannot be opened 180°. The hangers (Fig. 2-16) are of two-piece, universal construction. There are numerous combinations of the two-piece hanger which enable the wireway to be installed in any manner. Holes in the hanger match holes in the wireway and allow the wireway cover to be opened a full 180° (for easy lay-in).

Closing plates (Fig. 2-17) are used to seal the ends of the wireway or fittings. Captive screws and two ears in the trough anchor the plate tightly in place. Knockouts in the closing plate allow extension of the wireway through the use of conduit, tubing, or cable. Elbows (Fig. 2-17) are available in angles of 22.5°, 45°, and 90° (for making bends or fittings around building contours) and attach to lengths (Fig. 2-14) through the connectors (Fig. 2-15). Telescopic fittings, or slip fittings (Fig. 2-17), provide a means for adjusting straight-length connections at varying distances from 0.5 to 11.5 in. Setscrews

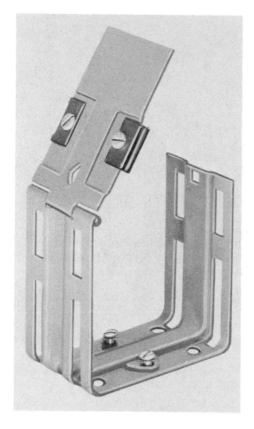

FIGURE 2–15 SQUARE-Duct wireway connector. (Courtesy Square D Company.)

provide a grounding connection between the sliding wireway sections. Adapter fittings (Fig. 2–17) are available for connecting the wireway to panelboards, switchboards, and cabinets.

Tee fittings (Fig. 2–18) are used for making tee connections on standard sections of wireway. Covers and sides are removable and are finished with self-aligning captive screws. Tee fittings are attached to the wireway with standard connectors. Junction boxes, or crosses (Fig. 2–18), have four openings and are coupled with regular connectors. Screw covers and slides retain the lay-in feature. Unused openings can be sealed with closing plates. Transposition sections permit the wireway opening to be rotated 90° and still retain the lay-in feature. Gusset brackets are used when mounting SQUARE-Duct from a wall. Hangers are not required when gusset brackets are used. Pull boxes are used when major splicing is necessary at a junction of several wireway runs. Pull boxes have six openings. Unused openings can be sealed with closing plates. Reducer fittings (Fig. 2–18) are available for joining many sizes of SQUARE-Duct.

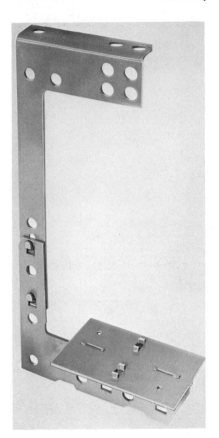

FIGURE 2-16 SQUARE-Duct wireway hanger. (Courtesy Square D Company.)

2-7.3 *Raintight Wireway*

Raintight wireway is a steel-enclosed wiring trough designed to be used outdoors, and in other areas where raintight construction is required. Raintight has a removable cover with provisions for sealing and is UL listed for ganging meters, switches, and other equipment suited for outdoor use.

2-7.4 *JIC Lay-in Wireway*

JIC lay-in wireway is a gasketed, metal wiring trough manufactured to meet specifications for Industrial Control Equipment, and is used to protect electrical wiring from oil, water, dirt, or dust. A complete line of elbows, tees, and other fittings fully complement the JIC line to meet all layouts. JIC wireway is manufactured with a minimum number of captive parts for quick assembly. Hex screws with slotted heads permit easy assembly, requiring only the use of hand tools. Each length comes complete with a connector kit and

(a)

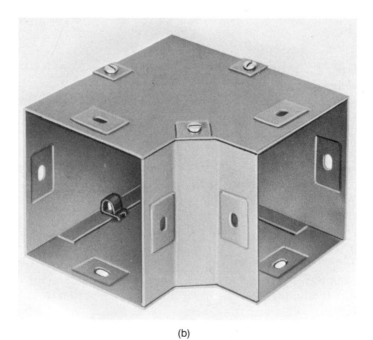

(b)

FIGURE 2-17   SQUARE-Duct wireway fittings, elbows, and closing plates. (Courtesy Square D Company.)

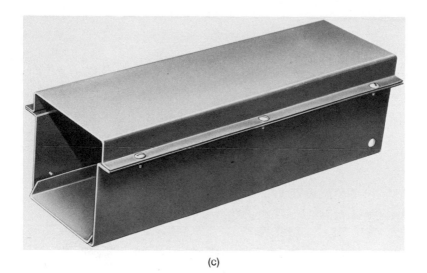

(c)

(d)

FIGURE 2–17 (*Cont.*)

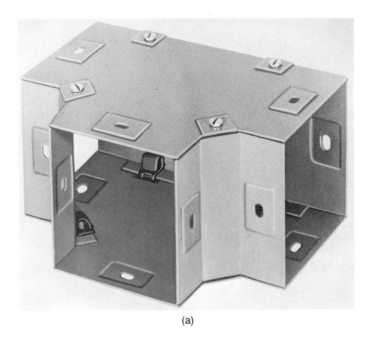

(a)

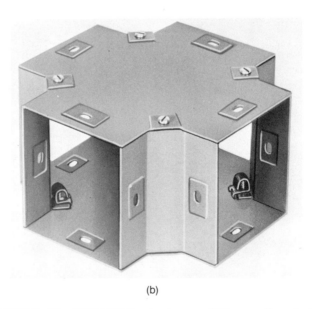

(b)

FIGURE 2-18   SQUARE-Duct wireway fittings, elbows, and junction box. (Courtesy Square D Company.)

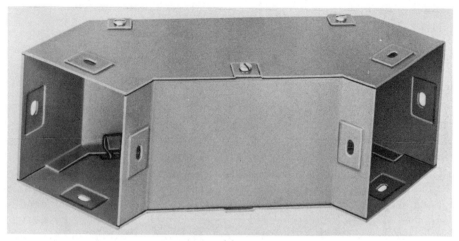

(c)

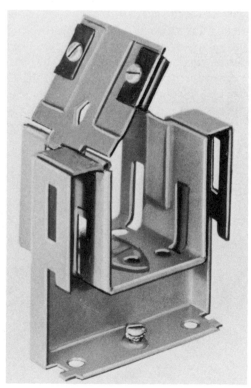

(d)

FIGURE 2-18 *(Cont.)*

hardware for joint connection. JIC wireway is UL listed as wireways and auxiliary gutters.

2-7.5   *Underfloor Duct*

Underfloor duct is a raceway system design to be embedded in the concrete floor of offices, classrooms, laboratories, manufacturing areas, supermarkets, and so on, for the purpose of providing an enclosed raceway for wires and cables from their originating panel or closest to their point of use. An underfloor raceway system is composed of two types of ducts: feeder ducts and distribution ducts. Complementing these two types of ducts are junction boxes, support couplers, supports, horizontal and vertical elbows, power and telephone outlets, and numerous cast and sheet-metal fittings used for conduit adapters, change of direction of duct runs, takeoffs, and so on. *Underfloor duct conforms to Article 354 of the NEC, as do other conduits and raceways.*

Figure 2-19 shows two underfloor ducts, both of which are *distribution ducts*. Such distribution ducts provide a distribution raceway from a junction with the feed raceway to the point of service. Distribution ducts have a fabricated insert every 24 in. along the length. These inserts provide for installation of service fittings (Fig. 2-20). The underfloor duct system also includes *feeder ducts,* which are similar to the distribution ducts but without the inserts. Feeder ducts provide the feed from the service terminal points (lighting panelboards, telephone closet or cabinet, signal cabinet, etc.) to the distribution duct.

Figure 2-21 shows the junction boxes used with the underfloor duct sys-

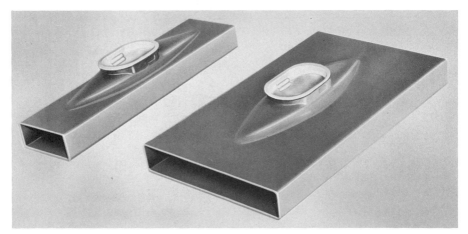

FIGURE 2-19 Underfloor distribution ducts. (Courtesy Square D Company.)

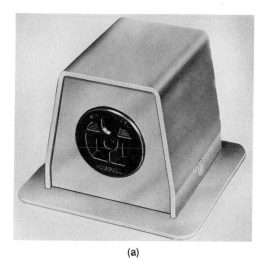

(a)

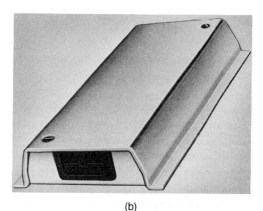

FIGURE 2-20 Service fittings for underfloor ducts. (Courtesy Square D Company.)

(b)

tem. Such junction boxes are used with the feeder and distribution ducts. The junction boxes have interior partitions (or tunnels) to maintain the separation of services where two or more noncompatible services form a junction. For example, in a two-duct system, where one raceway is for telephone cables and one raceway for branch-circuit wiring, the junction box provides a separate raceway for each system, both through the box and at right angles.

All Square D underfloor junction boxes have two means of leveling. The first adjustment is made using the threaded leveling legs shown in Fig. 2–21. These legs adjust the top of the box to the screed line of the concrete. The second method of leveling is a fine adjustment composed of four screws in the corners of the boxes. By using these four screws, the box can be adjusted after the concrete is poured.

FIGURE 2-21   Junction boxes for underfloor ducts. (Courtesy Square D Company.)

2-7.6   *Commercial and Industrial Applications for Wireways*

Square D wireway can be used in a variety of ways and, in many cases, is less expensive, and can be installed faster than conduit. Typical commercial and industrial applications where wireway is ideal include incoming cable runs and feeder circuits from switchboards to power and lighting panels; distribution of power in industrial plants, ganging equipment, such as motor control, safety switches, and metering equipment; and vertical runs of cable in elevator shafts. When comparing wireway with conduit, wireway has distinct advantages, including lighter weight, easier installation, reusability, and the ease of making tap-offs.

2-7.7   *Wireway Design Limitations*

All of the conductor design factors described in Secs. 2–1 through 2–4 apply to wireways, with the following major exceptions or limitations:

The number of conductors used in a wireway run is limited to not more than 30 at any cross section, unless the conductors are for signal currents, or are controller conductors between a motor and motor-starter, used only for starting duty.

The conductor cross-sectional area is limited to 20% of the wireway cross-sectional area.

In theory, conductors used in a wireway do not have to be derated as is the case with conduit. However, the author recommends that the conductors be designed as described in Secs. 2–1 through 2–4, including any derating factors. Then simply check that no more than 30 conductors are used in any wireway design, and that the cross-sectional area of the conductors does not exceed 20% of the wireway cross-sectional area. An example of selecting the correct wireway size is given in Sec. 2–9.

2-8 BUSWAYS

As discussed, bus bars are used in commercial and industrial installations when heavy currents are involved (typically, when the current exceeds about 500 A). Busways are essentially metal ducts with built-in bus bars, and opening at various points along the duct (permitting switches and outlets to be plugged into the bus bars as needed). With busways, you can add outlets, switches, circuit breakers, and so on, with a minimum of rework. Again, we make no effort to describe all types of busways here. Instead, we concentrate on busways manufactured by Square D Company, known as the I-LINE® busway.

Figures 2–22 and 2–23 show the I-LINE plug-in and feeder busway construction, respectively. I-LINE feeder busway is available in ratings from 600 through 4000 A (aluminum), and from 800 through 5000 A (copper), and is constructed in three-pole or four-pole full neutral for system voltages up to 600 V ac. I-LINE busway may carry two-pole ratings with UL listings up to 8650 A dc, or single-phase ac.

2-8.1 Enclosed Construction

Both I-LINE feeder and plug-in busways are totally enclosed and do not require ventilation openings in the housing for cooling. Instead, I-LINE busway cools by radiation from the housing surface and by convection currents out-

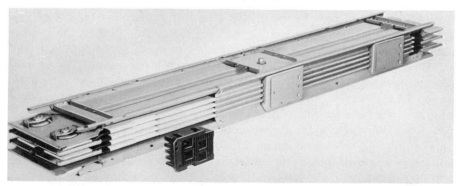

FIGURE 2-22 Construction of I-LINE plug-in busway. (Courtesy Square D Company.)

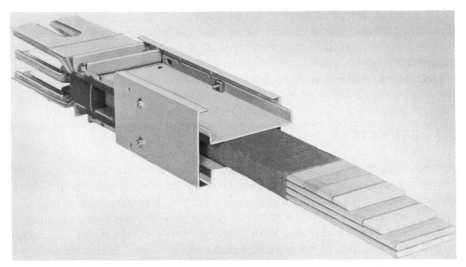

FIGURE 2-23 Construction of I-LINE feeder busway. (Courtesy Square D Company.)

side the housing. This method of cooling offers several advantages over conventional ventilated busways.

Conventional ventilated busways maintain maximum operating within allowable limits by using convection currents of air which pass through the housing to carry off excess heat. When ventilated busway is mounted so that the perforated housing allows these convection currents to pass freely through the housing and between the conductors, cooling is fairly efficient. However, if the busway is mounted in any position other than this preferred position, the bus bars interfere with free passage of cooling air, efficiency is decreased, and the operating temperature rises. Under these conditions, ventilated busway must be derated to a substantially lower current-carrying capacity (or oversized bus bars must be used). The temperature derating factors of Fig. 2-8 apply in the absence of specific derating factors supplied by the busway manufacturer. Always use the manufacturer's derating factors on busways (and any other electrical equipment). Where totally enclosed construction is used on busways that are normally ventilated, even more stringent derating is required. I-LINE busways need no derating at different mounting positions since cooling is the same for all positions.

The preferred mounting position of most ventilated busways requires the plug-in units (switches and circuit breakers) to be mounted on the top and bottom of the run, making those on top hard to get at, and those on the bottom protrude into available headroom. I-LINE busway plug-in units may be side-mounted for maximum use, without derating the busway.

The NEC requires that busway may extend vertically through dry floors if totally enclosed (unventilated), where passing through, and for a minimum distance of 6 ft above the floor to provide adequate protection from physical damage.

I-LINE busway complies with this NEC requirement without modification. In the case of ventilated busway, if the enclosure is not provided by a busway manufacturer, the busway may meet the requirements of NEC, but void the UL approval, since UL cannot sanction modifications made to a product in the field. Ventilated busways require extensive modification to satisfy both UL and NEC requirements. I-LINE busway requires no modification.

Totally enclosed construction provides much greater protection from mechanical damage to the bus bars and insulation. Also, enclosed construction gives much better protection from dust and dirt accumulation in the housing than does ventilated housing. Because I-LINE construction does not require that bus bars be spaced apart for airflow between them, the physical size of the housing can, with proper design, be smaller, rating for rating. Weight is also lower in I-LINE than in other types of busway.

The close spacing of bus bars in I-LINE busway results in very low *reactance* (Chapter 9). This is particularly true of feeder busway, where the spacing between bars in less than $\frac{1}{16}$ in. The low reactance helps reduce voltage dips during the instant of a change in load, such as motor starting (Chapter 8). Under such conditions, the high inrush current (up to 600% of load) is at a very low power factor (Chapter 9), and the reactive component of voltage drop assumes greater importance. The very low reactance of the I-LINE busway reduces the voltage dips. This is very important where voltage-sensitive equipment (X-ray machines, electronic computers) is in use, or where lamp flicker is objectionable.

2-8.2 *Short-Circuit Ratings*

Busways are rated as to their ability to withstand the effects of short circuits. When short circuits occur, bus bars are subjected to considerable stress. Some means of restraining the stress forces must be provided to prevent physical damage to the busway. The term *short-circuit bracing* is often used to describe the restraining function. Such short-circuit bracing is needed during current surges which result from short circuits or other low-resistance faults, either in the busway or in the equipment fed by the busway. Magnetic fields are set up around bus bars when current passes through the bars, and cause the bars to be repelled from each other (or drawn together). If the currents are heavy, the bars or the housing can be damaged. In a large busway system, the stress can reach values of several tons per lineal foot of conductor. Also, for a three-phase system which often uses busways for distribution, there is always one conductor being forced away from the other two conductors by fault current forces.

I-LINE busways provide short-circuit bracing by means of "epoxilated" (epoxy encapsulated) insulation. The wrapped insulated bars are assembled in the steel and aluminum housing while the epoxy is still wet. As the epoxy sets, the conductor "sandwich" is bonded to the housing to ensure good heat transfer and to confine short-circuit forces over the entire length of the bus bars. Further short-circuit bracing is provided by clamps and special molding insulators. All I-LINE feeder busway carries a conservative short-circuit rating of 100,000, 150,000, or 200,000 rms symmetrical amperes, depending on the size of the busway.

The support provided at the plug-in area is particularly important. Without such support, movement of the bus bars may damage plug-in units installed on the busway, even though the busway is able to withstand the fault and shows no apparent damage. This is why NEMA (National Electrical Manufacturers Association) Standards BU 1–3.03 and BU 1–3.04 require that the short-circuit rating of any plug-in busway be arrived at by testing two lengths of busway in series, one of which has plug-in units installed and one without plug-ins.

2-8.3   *Plug-in Busway and Units*

I-LINE plug-in busway (Fig. 2–22) provides a tap off every 2 ft along both sides of the length. This makes 10 openings per 10-ft length. Plug-in opening spacing is not disturbed by joint location on lengths of even footage. All openings are usable. Plug-in busways with ratings of 800 A and larger use a swing-away base feature which allows bolt-on-style units up to 1600 A capacity to be connected at any plug-in opening. All plug-in units are provided with a grounding spring which makes a positive ground connection between the plug body and the busway housing, prior to jaw contact with the bus bars. Also, the manner in which plug-in units are mounted polarizes the units so that the neutral is always the bottom conductor.

The Square D catalog provides full information concerning the plug-in units available for use with the I-LINE busway. The following is a brief summary:

Fusible units from 30 to 1600 A

Circuit breaker units from 15 to 1600 A

Combination starter and contactor units through NEMA size 3

Transformer units through 10 kva

Capacitor units through 30 kvar

Ground detector units

2-8.4   Feeder Busway

I-LINE feeder busway (Fig. 2–23) is manufactured in weatherproof as well as indoor construction. The weatherproof design incorporates gasketed covers for joint parts, vapor barriers, and other features which make it possible to install weatherproof busway in exposed locations. Weatherproof busway can be connected to indoor feeder or to plug-in feeder busway with a standard joint.

Plug-in busway is manufactured only in the indoor configuration and *should not be used outdoors.* No feeder or plug-in busway is suitable for extremely dusty or hazardous locations, or in extremely corrosive atmospheres. However, I–LINE busway, because of its totally enclosed housing, full insulation, and corrosion-resistant tin plate, is often more suitable for borderline cases than other types of busway.

2-9   DETERMINING RACEWAY SIZE

Although the type of raceway may be set by local code, or by cost factors, or by the ampacity required, *the size of the raceway is determined by NEC requirements.* In the case of cables (either metal-clad or nonmetallic) where the conductors are manufactured as one piece with the outer shield, there is no need to calculate the raceway size. Also, the manufacturers of busway set the number of conductors (usually four), and rate the busways as to ampacity and operating voltage limits. However, it is necessary to determine raceway size for rigid conduit and EMT. Also, the correct size must be selected for both wireways and busways. Calculations for determining raceway size are covered in the remaining paragraphs of this section.

2-9.1   Calculating Rigid Conduit and EMT Size

Figure 2–24 provides a means for calculating the maximum number of conductors permitted by the NEC (for new work) to be used in various trade sizes of conduit tubing. The information in Fig. 2–24 can also be used to find raceway size.

As an example, assume that you have found that a particular installation requires four No. 0000 AWG conductors to get the required minimum ampacity and maximum voltage drop (using the calculations of Secs. 2–1 through 2–4). Now it is necessary to select the correct size for rigid conduit or EMT raceway. Start by finding the smallest size of raceway permitted.

Using the No. 0000 AWG size, move vertically down the row until the first number 4 is found. If you are paying attention, you will see that there is no number 4 in the No. 0000 AWG row. The numbers change from 3 (for $2\frac{1}{2}$-in. raceway) to 6 (for 3-in. raceway). So the smallest permitted raceway

Conduit size (inches)	14	12	10	8	6	4	3	2	1	0	2/0	3/0	4/0	250	300	350	400	500	600	700	750	800	900	1000
1/2	4	3	1	1	1	1																		
THWN	8	6	4	2	1	1																		
3/4	6	5	4	3	1	1	1	1	1															
THWN	15	11	7	4	2	1	1	1	1															
1	10	8	7	4	3	1	1	1	1	1	1	1												
THWN	24	18	11	6	4	2	2	1	1	1	1													
1-1/4	18	15	13	7	4	3	3	3	1	1	1	1	1	1	1	1								
THWN	43	32	20	11	7	4	3	3	2	2	1	1	1	1	1	1								
1-1/2	25	21	17	10	6	5	4	3	3	2	1	1	1	1	1	1	1	1						
THWN	58	43	27	16	9	6	5	4	3	2	2	1	1	1	1	1								
2	41	34	29	17	10	8	7	6	4	4	3	3	2	1	1	1	1	1	1	1	1	1	1	1
THWN	96	71	45	26	16	9	8	7	5	4	3	3	2	2	1	1	1	1	1	1	1	1	1	1
2-1/2	58	50	41	25	15	12	10	9	7	6	5	4	3	3	3	1	1	1	1	1	1	1	1	1
THWN	137	102	65	37	23	14	12	10	7	6	5	4	3	3	3	2	2	1	1	1	1	1	1	1
3	90	76	64	38	23	18	16	14	10	9	8	7	6	5	4	3	3	3	1	1	1	1	1	1
THWN	—	158	100	58	35	21	18	15	11	9	8	7	6	5	4	3	3	3	2	2	1	1	1	1
3-1/2	121	103	86	52	32	24	21	19	14	12	11	9	8	6	5	5	4	4	3	3	3	2	1	1
THWN	—	—	134	78	47	29	24	20	15	13	11	9	8	6	5	5	4	4	3	3	3	2	2	1
4	155	132	110	67	41	31	28	24	18	16	14	12	10	8	7	6	6	5	4	3	3	3	3	3
THWN	—	—	172	100	61	37	31	26	20	16	14	12	10	8	7	6	6	5	4	3	3	3	3	3
4-1/2	197	168	140	85	52	40	35	31	23	20	18	15	13	11	9	8	7	6	5	4	4	4	4	3
THWN	—	—	—	127	78	48	40	34	25	21	18	15	13	11	9	8	7	6	5	4	4	4	4	3
5	—	—	—	173	105	64	49	44	38	29	25	22	19	16	13	11	10	9	8	6	6	5	5	4
THWN	—	—	—	—	157	96	59	50	42	31	26	22	19	16	13	11	10	9	8	6	6	5	5	4
6	—	—	—	152	93	72	63	55	42	37	32	27	23	19	16	15	13	11	9	8	8	7	7	6
THWN	—	—	—	—	139	85	72	61	45	38	32	27	23	19	16	15	13	11	9	8	8	7	7	6

FIGURE 2-24 NEC requirements for new work, maximum number of conductors in conduit.

size for four to six conductors is 3 in. This is true even if THWN nylon insulation is used on the conductors. For some conductor sizes, additional conductors can be used in the same size raceway if the insulation is different. For example, if you can use No. 3 AWG conductors in a 3-in. raceway, you can use 16 standard conductors, or 18 conductors with THWN insulation.

Depending on cost and other factors, it is sometimes convenient to use raceways that are larger than required. This permits a greater number of conductors, or conductors of a larger size, to be used at a later date (for remodeling or whatever).

For example, assume that you select $3\frac{1}{2}$-in. raceway instead of the 3-in. raceway described in our example. With $3\frac{1}{2}$-in raceway, eight No. 0000 AWG conductors can be used instead of six. This permits two four-wire systems to be housed in the same raceway. As an alternative, with a $3\frac{1}{2}$-in. raceway, the conductor size can be increased to 500 MCM for a single four-wire system (almost doubling the ampacity from that of No. 0000 AWG).

2-9.2   Calculating Wireway Size

Now let us assume that you want to use the SQUARE-Duct wireway (Sec. 2-7) instead of conduit to accommodate the single four-wire system with No. 0000 AWG conductors.

Obviously, the limit of 30 conductors in any cross section of a wireway presents no problem since there are only four conductors in our example. So the only problem is that the cross-sectional area of the conductors does not exceed 20% of the wireway cross-sectional area.

Start by finding the total cross-sectional area of the conductors. As shown in Fig. 2–2, the circular mil area of No. 0000 AWG is 211,600, which, when multiplied by 4, produces a total cmil area of 846,400. Since wireways are square, you must find the square-mil equivalent of 846,400 using the equations of Sec. 2–1.2, or 846,400 × 0.7854 = 664,263.

Now multiply the conductor area by 5 to get a total of 3,321,315. Next, *find the smallest* SQUARE-Duct *wireway* that has a square-mil area greater than 3,321,315. As shown in Fig. 2–14, the smallest SQUARE-Duct wireway size is $2\frac{1}{2} \times 2\frac{1}{2}$ in., which as a square-mil area of 6,500,000 (2500 × 2500). So you can use $2\frac{1}{2} \times 2\frac{1}{2}$ SQUARE-Duct to accommodate the four-wire system with four No. 0000 AWG conductors. If you are interested, divide the 664,263-square mil area of the conductors by the 6,500,000 area of the wireway to find that the conductor area is about 10.2% of the wireway, well below the 20% limit.

2-9.3 Calculating Busway Size

Now assume that you want to use busway instead of conduit or wireway to accommodate our four-wire system. From a practical standpoint, this is not likely since the maximum current of No. 0000 AWG conductors is on the order of 200 A (when derated for four conductors in a raceway), and a busway is generally most effective when currents exceed 500 A. However, let us assume that you anticipate an increase in the ampacity requirements for the future, and that you want to use the I-LINE busway described in Sec. 2–8.

Since the I-LINE busway is available in four-conductor, there is no problem with the number of conductors. Also, since busway is rated as to ampacity and operating voltage by the manufacturer, it is only necessary to make sure that the busway size selected accommodates the required ampacity and voltage. However, as a final check, it is necessary to determine the voltage drop produced by the busway length. This can be done using the calculations of Sec. 2–4, and the *published resistance* of the busway (which is usually given in terms of ohms per 1000 ft). If the voltage drop produced by the required length of busway exceeds the NEC maximums, even though the busway ampacity is sufficient to accommodate the load, you must use a larger busway size (with lower resistance for a given length, and possibly with much more than enough ampacity).

As an example, assume that busway is to be used for a 240-V single-phase system with a current of 500 A. Further assume that the total permitted voltage drop at the load is 5% of the source voltage. Under these conditions, the maximum allowable voltage drop is 12 V (240 V × 0.05). Further assume that the published resistance for a 600-A 600-V busway is 0.01 Ω per 1000 ft. The distance from the service entrance to the load (including feeders) is 1000 ft.

To find the maximum voltage drop, multiply the distance between the service and load by 2 (1000 ft × 2 = 2000 ft), then multiply the result by the published resistance to find the total resistance (0.01 Ω per 1000 ft × 2 = 0.02 Ω). Now multiply the total resistance by the current to find the actual voltage drop (0.02 Ω × 500 A = 10 V). This is below the maximum permitted voltage drop of 12 V.

Of course, this example assumes a simple installation with a single 500-A load located 1000 ft from the service entrance, using single-phase. Many practical busway applications have multiple loads, at different distances, requiring both branch and feeder circuits, and possibly three-phase. For that reason, all of Chapter 4 is devoted to design of branch and feeder circuits and applies whether the conductors are wires (in conduit or wireways) or bus bars (in busways).

3

CIRCUIT BREAKERS,
SAFETY SWITCHES, AND FUSES

Virtually all commercial and industrial wiring installations include at least one (and usually many more) circuit breakers and safety switches. Fuses are also used, particularly where short-circuit currents are very high, although fuses have generally been replaced by circuit breakers where anticipated short-circuit currents are less than 100,000 A.

It is essential that you understand breaker, switch, and fuse characteristics to design any wiring system properly. In the simplest of terms, you must know what is available, how the equipment is rated, what the ratings mean, and how the ratings affect design of a wiring system. Once you have digested this information, you can go on to the calculations for determining breaker, switch, and fuse sizes or capacities for service entrances, panelboards, load centers, feeder and branch circuits, and so on. Such calculations are described in Chapter 4.

In this chapter we describe the fuse classifications most commonly found in commercial and industrial wiring, but we make no attempt to cover every type and classification of breaker and switch. Instead, we concentrate on the breakers and switches manufactured by the Square D Company, since their line represents a cross-section of equipment now available for any wiring system. Note that there is a further discussion of circuit breakers and switches in Chapter 4 (as they apply to feeder and branch-circuit overprotection) and in Chapters 7 and 8 (as they apply to lighting and motors).

3-1 DEFINITION OF A CIRCUIT BREAKER

Figure 3-1 shows some typical circuit breakers suitable for commercial and industrial wiring systems. Such circuit breakers provide for closing and interrupting a circuit between separable contacts under both normal and abnormal conditions. This is done manually (normal condition) by means of the handle, which is switched to the ON or OFF position, as required. The circuit breaker is also designed to open a circuit automatically on a predetermined overload or short-circuit current (abnormal condition) without damage to itself or the associated equipment. The phenolic molded-case housing of the circuit breaker insulates all live parts, thus protecting the associated equipment on the circuit, as well as operating personnel, from accidental shock.

3-1.1 *Circuit Breaker Components or Parts*

Figure 3-2 shows the basic electrical circuit for the internal arrangement of a circuit breaker. A standard molded-case circuit breaker usually contains (1) a set of contacts, (2) a magnetic trip element, (3) a thermal trip element, (4) line and load terminals, (5) busing used to connect these individual parts, and (6) housing of insulating material.

  The circuit breaker handle manually opens and closes the contacts and resets the automatic trip units after an interruption. Square D circuit breakers also contain a manually operated "push-to-trip" testing mechanism. Each of these elements is discussed in the following paragraphs.

FIGURE 3-1 Typical circuit breakers suitable for commercial/industrial wiring systems. (Courtesy Square D Company.)

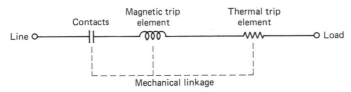

FIGURE 3-2 Basic electrical circuit for the internal arrangement of a circuit breaker.

3-1.2 Circuit Breaker Groupings

Circuit breakers are grouped for identification according to given current ranges. Each group is classified by the *largest* ampere rating of the range. For example, typical groups are 15–100 A, 125–225 A, 250–400 A, 500–1000 A, and 1200–2000 A. These groups are classified as 100-, 225-, 400-, 1000-, and 2000-A frames. These classification numbers are commonly referred to as *frame classifications* or *frame sizes* and are the terms applied to groups of molded-case circuit breakers which are physically interchangeable with each other. Square D circuit breakers are available in either standard or high-interrupting ratings (Sec. 3-2) in each frame size.

3-2 CIRCUIT BREAKER RATINGS

In addition to frame sizes, circuit breakers are rated primarily as to voltage, current, and interrupt capacity. The basic NEC and UL ratings for circuit breakers are as follows.

3-2.1 Circuit Breaker Voltage Ratings

The voltage rating of a circuit breaker is based on the clearances or space (both through air and over surfaces) between all components of the electrical circuit, and between the electrical components and ground. Circuit breaker voltage ratings indicate the maximum electrical system voltage on which they can be applied. The UL recognizes only the ratings listed in Fig. 3-3.

A circuit breaker can be rated for either a-c or d-c system applications, or both. Single-pole circuit breakers, rated at 120/240 V ac, or 125/250 V dc, can be used singly and in pairs on three-wire circuits having a neutral connected to the midpoint of the load, or used in pairs on a two-wire circuit connected to the outside (ungrounded) wires of a three-wire system. Two-pole or three-pole circuit breakers rated 120/240 V ac, or 125/250 V dc, can be used *only* on a three-wire dc, or single-phase ac, system having a grounded neutral (Chapter 5).

Alternating current (V)	Direct current (V)
120	125
120/240	125/250
240	250
277	600
277/480	
480	
600	

FIGURE 3-3 Circuit breaker voltage ratings.

Circuit breaker voltage ratings must be equal to or greater than the voltage of the electrical system on which the breakers are used. Figure 3–4 shows typical Square D circuit breakers and their group classifications.

3-2.2 Circuit Breaker Current Ratings

Circuit breakers have two types of current ratings. The first, and the one that is used most often, is the *continuous current* rating. The second is the short-circuit current *interrupting capacity.*

The rated *continuous current* of a circuit breaker is the maximum current in amperes that the breaker will carry continuously without exceeding the specified limits of observable temperature rise. Continuous current ratings of circuit breakers are based on standard UL ampere ratings, which are: 15, 20, 25, 30, 35, 40, 45, 50, 60, 70, 80, 90, 100, 110, 125, 150, 175, 200, 225, 250, 300, 350, 400, 450, 500, 600, 700, 800, 1000, 1200, 1600, 2000, 2500, 3000, 4000, 5000, and 6000 A.

The ampere rating of a circuit breaker is located on the handle, and the numerical value alone is shown (150 at the tip of the handle indicates 150 A). *The NEC requires that the circuit breaker current rating be equal to or less than the load circuit conductor ampacity.*

The *interrupting capacity* rating (known as the AIC rating) of a circuit breaker is the maximum short-circuit current that the breaker will interrupt safely. This AIC rating is at rated voltage and current. Where the breaker can be used on more than one voltage, the interrupting capacity is shown for each voltage level. For example, the LA-type circuit breaker (Fig. 3–4) has 42,000 A (symmetrical) interrupting capacity at 240 V, 30,000 A symmetrical at 480 V, and 22,000 A symmetrical at 600 V. (Compare these values to those of Fig. 1–9.)

Note that the term "symmetrical" as applied here refers to the waveform of the current. The electrical currents generated by most utility companies are considered to be sine waves and are symmetrical. Virtually all voltage and current ratings of electrical equipment manufacturers are based on the as-

FIGURE 3-4   Typical Square D circuit breakers.

Catalog number prefix	Number of poles	Maximum a-c voltage rating	Ampere rating
QO-GFI, QOB-GFI	1	120	15–30
	2	120/240	15–30
QO-VHGFI, QOB-VHGFI	1	120	15–20
QO-QOR	1	120/240	10–70
	2	120/240	10–70
	3	240	10–60
QO-H, QOB-H	2	240	15–30
QOT	1	120/240	15–30
	2	120/240	15–30
QO-VH, QOB-VH	1	120/240	10–30
	2	120/240	15–30
	3	240	15–30
QH-QHB	1	120/240	10–30
	2	120/240	15–30
	3	240	15–30
Q1-Q1B	1	120/240	80–100
	2	120/240	110–150[a]
	3	240	70–100
Q1-H, Q1B-H	2	240	35–100
Q1-VH, Q1B-VH	2	120/240	35–125[b]
	3	240	35–100
Q1H, Q1HB	2	120/240	35–125[b]
QOU	1	120/240	10–50
	2	120/240	10–70
	3	240	10–50
Q1L-Q1U	2	240	15–125
	3	240	15–100
Q2	2	120/240	100–225
Q2L	2	240	100–225
Q2-Q2L	3	240	100–225
Q2-H, Q2L-H	2	240	100–225
	3	240	100–225
Q4, Q4L	2	240	250–400
	3	240	250–400
EH, EHB	1	277	15–30
	2	227/480	15–60
	3	227/480	15–60
FY	1	277	15–30
FA-FAL	1	120	15–100
	2	240	15–100
	3	240	15–100
FA-FAL	1	277	15–100
	2	480	15–100
	3	480	15–100
FA-FAL	2	600	15–100
	3	600	15–100
FH-FHL	1	277	15–30
	1	277	35–100
	2	600	15–100
	3	600	15–100

FIGURE 3-4   *(Cont.)*

Catalog number prefix	Number of poles	Maximum a-c voltage rating	Ampere rating
IF-IFL	2	480	15–100
	3	480	15–100
KA-KAL	2	600	70–225
	3	600	70–225
KH-KHL	2	600	70–225
	3	600	70–225
IK, IKL	2	480	110–225
	3	480	110–225
LA-LAL	2	600	125–400
	3	600	125–400
LH-LHL	2	600	125–400
	3	600	125–400
MA-MAL	2	600	125–1000
	3	600	125–1000
MH-MHL	2	600	125–1000
	3	600	125–1000
NHL	2	480	600–1200
	3	480	600–1200
PAF	2	600	600–2000
	3	600	600–2000
PHF	2	600	600–2000
	3	600	600–2000
PEF, PEC	2	600	1000–2000
	3	600	1000–2000
PCF	2	600	1600–2500
	3	600	1600–2500

[a] Q1 only.

[b] Q1B–VH and Q1HB are 100 A maximum.

sumption that the current is symmetrical (or other ratings are specified for different types of currents). For example, if a circuit breaker is used with a system where the currents are pulsed (where the waveform has sharp peaks and is not symmetrical), the current ratings must be reduced from those specified for symmetrical waveforms.

3-2.3 Circuit Breaker Interrupting Capacity Ratings

Circuit breakers are available with both standard and high interrupting capacities. Circuit breakers with *standard interrupting capacities* are identified by a *black operating handle* and *black printed interrupting rating labels,* and have an AIC of 10,000 A. *High interrupting capacity breakers* are identified by a *gray handle* and *red interrupting rating labels,* and have an AIC of up to 65,000 A. Square D high-interrupting-capacity breakers are identified as an I-75,000-type breaker.

3-2.4 Circuit Breaker Frequency Ratings

The standard rated frequency for circuit breakers is 60 Hz. Frequencies below 60 Hz usually do not affect the trip setting (Sec. 3-4). However, molded-case circuit breakers used on systems with frequencies of 25 Hz or less should be selected on the basis of the d-c ratings.

The overcurrent trip rating (thermal trip elements, Sec. 3-4) in Square D Q2, KA, LA, and MA circuit breakers should be derated to approximately 80% of the 60-Hz value when used on 400-Hz systems. However, the short circuit (magnetic trip rating) is not affected. The manufacturer should be consulted if a circuit breaker is used on any frequency above 400 Hz.

3-2.5 Circuit Breaker Ambient Temperature Ratings

The rated ambient temperature of a circuit breaker is the temperature on which the continuous current rating (not the interrupt capacity) is based. Ambient temperature is the temperature of the air immediately around the circuit breaker and can affect the tripping point. Generally, ambient temperature is considered to be 25°C for wiring design. If circuit breakers are to be operated in high-temperature areas, the breakers should be derated. Always consult the manufacturer's specifications for any such derating.

3-2.6 How Circuit Breaker Ratings
Affect Wiring Design

The calculations for selecting circuit breakers with the correct continuous current rating are relatively simple. This is not true when selecting circuit breakers with the correct interrupting capacity.

Continuous Current. Obviously, a circuit breaker *cannot* have a continuous current ampacity greater than the conductor being protected. Otherwise, the conductor can be damaged without tripping the circuit breaker. Also, the circuit breaker *must have* a continuous current ampacity equal to (or greater than) the anticipated load. Otherwise, the circuit breaker will trip when the load is applied. As discussed in Chapter 4, this is no particular problem if conductors of the correct ampacity are selected. Generally, the problem is solved by noting the load ampacity and selecting the *next largest* standard-size circuit breaker. For example, if the load ampacity is 90 A, use a breaker with a 100-A continuous current rating. Then check that the conductors are capable of handling 100 A.

Interrupting Capacity. The calculations for overcurrent protection in both feeders and branch circuits (Chapter 4) is based on simple direct current, when considering continuous current ratings. On the other hand, calculations for interrupt capacity are generally based on alternating current. This is because interrupt capacity is based on *available short-circuit current*. In turn, short-circuit current is based on voltage and impedance (for ac) or voltage and resistance (for dc). The impedance at any point in a wiring distribution is much greater than the simple d-c resistance at that same point. (Refer to Chapter 9 if you are not familiar with the terms "impedance" and "resistance.") If direct current is used to calculate the available short-circuit current, the interrupt capacity ratings are unnecessarily high (resulting in more expensive circuit breakers).

Unfortunately, the calculations for available short-circuit current are very complex when ac is used. As covered in Chapter 9, impedance is a combination of reactance (X) and resistance (R), and is calculated by the ratios of X and R. Although the resistance of a conductor is relatively constant (for a given material, size, etc.), the reactance depends on many factors, such as the number of conductors in a raceway, raceway material (magnetic or nonmagnetic), and frequency. The problem is further complicated if three-phase is involved. When there is a short on one phase of three-phase, the system appears to be single-phase being driven by three-phase voltages. This is known as the *single-phase effect*. Also, if a short occurs on any wiring system, there is a temperature rise, which produces changes in resistance but not in reactance.

One solution to this problem is to calculate the available short-circuit current on the basis of dc, and then assume that the a-c short-circuit current is no more than 80% of the d-c value. This system is used in Chapter 4.

An even more practical, and safe, solution to the available short-circuit problem is to use the *current-limiting* circuit breakers described in Sec. 3–3.

3-3   CURRENT-LIMITING CIRCUIT BREAKERS

Figure 3–5 shows two current-limiting circuit breakers (called I-LIMITERs® by Square D). In addition to performing the standard circuit breaker functions, I-LIMITERs limit the current to a certain level before they trip. This makes it possible to protect other breakers and wiring downstream (toward the load) from the I-LIMITERs in the event of heavy currents produced by short circuits or other faults. For example, assume that a system protected by standard breakers designed to trip at 1000 A is subjected to a 100,000-A short circuit current. The 100,000-A current may damage the wiring or breaker before the 1000-A breaker trips.

The need for current limitation is the result of increasingly higher available short-circuit currents associated with the growth and interconnection of modern power systems, particularly industrial systems. Before current-limiting breakers were developed, the only practical downstream protection for branch

FIGURE 3-5 Typical I-LIMITER current-limiting circuit breakers. (Courtesy Square D Company.)

circuits was to use fuses in the feeders. The I-LIMITER does not use fuses, so there is nothing to replace, even after clearing a maximum-level fault current.

The manufacturers of current-limiting circuit breakers often provide curves to show the limiting characteristics. For example, there are *peak let through curves* (I_p) that show the maximum current before the breakers are tripped, and there are $I^2 T$ *curves* that show the extent to which the breaker limits energy let-through under short-circuit conditions.

From a simplified design standpoint, the main point to remember is that current-limiting circuit breakers reduce short-circuit currents to some value below the interrupt capacity or AIC of other breakers in the system. This makes it possible to use less expensive breakers with a lower AIC in the system, and still have the necessary short-circuit protection.

One of the major uses of current-limiting circuit breakers is as mains in panelboards, to protect other circuit breakers in the system. Typically, the current-limiting circuit breakers are used in the feeders to protect standard breakers in the branch circuits. This results in a *total integrated wiring system* with a very high short-circuit interrupt capability. For example, when Square D I-LIMITER circuit breakers are used, an *integrated current interrupt rating* of 100,000 A can be obtained.

Figure 3-6 shows a label applied to Square D panelboards that include I-LIMITERs at the mains. From a practical wiring design standpoint, 100,000 A of current interrupt is generally sufficient for any application, including heavy industrial wiring systems. However, even on systems not requiring the full 100,000-A rating of the I-LIMITER main, considerable cost savings can

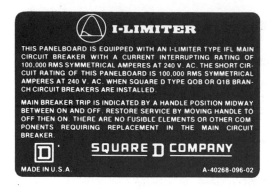

THIS PANELBOARD IS EQUIPPED WITH AN I-LIMITER TYPE IFL MAIN CIRCUIT BREAKER WITH A CURRENT INTERRUPTING RATING OF 100,000 RMS SYMMETRICAL AMPERES AT 240 V. AC. THE SHORT CIRCUIT RATING OF THIS PANELBOARD IS 100,000 RMS SYMMETRICAL AMPERES AT 240 V. AC. WHEN SQUARE D TYPE QOB OR Q1B BRANCH CIRCUIT BREAKERS ARE INSTALLED.

MAIN BREAKER TRIP IS INDICATED BY A HANDLE POSITION MIDWAY BETWEEN ON AND OFF. RESTORE SERVICE BY MOVING HANDLE TO OFF THEN ON. THERE ARE NO FUSIBLE ELEMENTS OR OTHER COMPONENTS REQUIRING REPLACEMENT IN THE MAIN CIRCUIT BREAKER.

FIGURE 3-6 Label applied to Square D panelboards that include I-LIMITERs as the mains. (Courtesy Square D Company.)

often be realized by using the I-LIMITER main and, for instance, FY breakers for the branch circuits on systems that would otherwise require all FH breakers for the branches. (Note that Square D FY breakers have standard interrupt capacities, while FH breakers have high interrupt capacities.)

Conductors that have high ampacity also have high available short-circuit currents. Thus busways have the greatest need for protection. Figure 3–7 shows I-LIMITER circuit breakers that can be used with the I-LINE busways described in Sec. 2–8. The I-LIMITER circuit breakers of Fig. 3–7 include plug-on connectors and a steel mounting bracket specifically designed to properly position and support the circuit breakers when mounted in I-LINE panelboards and switchboards (both of which are discussed in Chapter 4).

FIGURE 3-7 Typical I-LIMITER circuit breakers used with I-LINE busways. (Courtesy Square D Company.)

3-4 CIRCUIT BREAKER TRIP ELEMENTS

There are two basic types of overcurrent trip elements in circuit breakers: *thermal trip* (with built-in delay) and *magnetic trip* (for immediate trip).

3-4.1 Thermal Trip Element

An overcurrent trip element is a device which, for a given pole of a circuit breaker, detects overcurrent and transmits the energy necessary to trip the breaker automatically. The overcurrent trip element in Square D circuit breakers is commonly called a *thermal trip element,* meaning that the element is sensitive to heat. These elements have an *inverse time characteristic,* which indicates that a delayed action is purposely introduced, and that the delay decreases as the current increases.

Square D breakers rated 100 A and lower use *directly heated* overcurrent trip elements, where the load current passes through the element. Breakers rated 125 A and above have *indirectly heated* overcurrent elements. One end of the element is welded to the breaker busing, and the heat is transferred by conduction into the element.

Other types of overcurrent elements are the *shunted* and the *shorted turn.* The shunted element is heated directly by a portion of the current that passes through the breaker, while the busing carries the remainder of the current. The shorted-turn element is heated by transformer action and is good only for a-c applications.

Overcurrent trip elements are bimetal strips that bend as heat is applied. Current passing through the breaker creates most of the effective heat used for the tripping action. The ambient temperature surrounding the breaker either adds to or subtracts from this available heat. One of two ambient temperatures (25°C or 40°C) are used to calibrate the circuit breaker. Square D circuit breakers are calibrated at 40°C. A breaker carries the rated current continuously when the actual ambient air is at the same or lower temperature as the rated ambient temperature. The thermal overload trip element is permanently fixed at the factory in all circuit breakers, and cannot be adjusted in the field.

3-4.2 Magnetic Trip Element

The magnetic trip element contains an electromagnet which trips the breaker instantly at or above a predetermined value of current. All standard Square D circuit breakers have a magnetic trip element in each pole. The magnetic trip element is a U-shaped yoke and an armature arrangement that responds to a given value of current and is not affected by heat (as is the thermal element).

Both fixed and adjustable magnetic trip elements are available. Standard breakers rated 100 A and below have fixed magnetic trip elements. Breakers rated 125 A and above have adjustable magnetic trip elements (except for Type Q1 and Q2 circuit breakers). An exclusive feature of Square D circuit breakers is the *one control adjustment* of the magnetic trip. This one control adjusts all poles at the same time and to the same value as that of the tripping current. The adjustment is continuous from 5 to 10 times the breaker continuous current rating. The trips may be set at any value of current from the lowest to the highest edge of the trip range.

3-4.3 *Push-to-Trip*

Push-to-trip is an added feature of the Square D *industrial class* of circuit breaker. The push-to-trip button, shown in Fig. 3–8, permits the operator to trip the circuit breaker manually without exposing the operator to live parts in the breaker. Primarily, the push-to-trip feature is used to check for proper

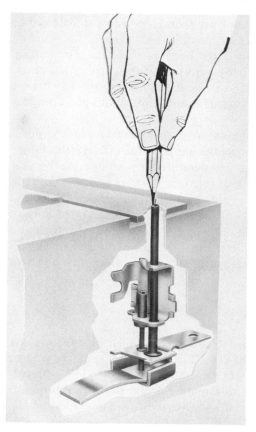

FIGURE 3-8 Push-to-trip mechanism on Square D circuit breakers. (Courtesy Square D Company.)

and safe operation of the tripping mechanism. Also, some manufacturers use the feature to adjust external handle operating mechanism. Maintenance personnel use the feature to check alarm circuits, emergency circuits, to train operators in motor sequencing operations, and to diagnose electrical problems. The push-to-trip feature assures the user that the breaker is in operable condition, since all operating parts are exercised when testing in this manner.

3-4.4 *Magnetic-Trip-Only Circuit Breakers*

Some circuit breakers are operated by a magnetic trip element only, and have no thermal trip unit. Such breakers trip instantaneously at or above a predetermined value of current. In the Square D line, such devices are called MAG-GARD® circuit breakers.

Magnetic-trip-only breakers are used in those applications where *only short-circuit protection* is desired. In such applications, the hazard due to lack of overcurrent (thermal tripping) protection is less objectionable than the outage of the circuit, except for the worst-case (short-circuit) condition. The magnetic-trip-only breaker is also used in those applications where a very high inrush current might cause nuisance tripping of the thermal trip unit. Typical applications which normally require magnetic-only breakers are fire pump hoses, welders, and so on.

Magnetic-trip-only breakers can carry their continuous current ratings indefinitely, but may be damaged by currents larger than continuous rating (even though smaller than the trip rating) if the currents are allowed to continue for long periods of time. The MAG-GARD circuit breaker is not available in all the continuous current ratings, as are the conventional thermal-magnetic breakers. However, overlap is provided between ratings to completely cover the entire range of values from 7 to 10,000 A.

3-5 CIRCUIT INTERRUPTERS

Circuit interrupters are not true circuit breakers, but are often considered to be the same. A circuit interrupter is essentially a switch in a molded case that is identical in size and appearance to those used with circuit breakers. Circuit interrupters are available in both automatic and nonautomatic types. The main purpose of a circuit interrupter is to meet *NEC article 430–102, which requires that a disconnecting means be placed within sight of all motor control locations.* (Motor control is discussed in Chapter 8.)

Nonautomatic circuit interrupters are used to meet the NEC requirement when the overcurrent protection is out of sight from the controller. Nonautomatic interrupters do not have any trip elements and are manually operated, but have standard breaker contacts, busing, and lugs for the highest ampere rating in each breaker frame size. Although the nonautomatic interrupter is used primarily for motor control, the interrupters need not be horsepower rated because the interrupting ability is much greater than the locked rotor currents produced on such circuits. Nonautomatic interrupters will withstand fault current levels between 15 and 20 times the continuous ratings. FA, KA, LA, MA, and PA nonautomatic circuit interrupters are available in the Square D line, but are not available in the I–75,000 type (except for the PE breaker).

Automatic circuit interrupters are used where the available fault current is greater than the nonautomatic interrupter will withstand, and are generally safer to use than nonautomatic interrupters. Automatic circuit interrupters have fixed magnetic trip elements set to actuate on current values of 20 times the corresponding circuit breaker rating (and above). The automatic circuit interrupter *is not* intended to be a circuit overcurrent protective device. Instead, a high setting of the trip element in the interrupter allows the feeder-to-branch circuit overcurrent protective device (typically a circuit breaker) to do its job without interference. The automatic circuit interrupter has the same interrupting ratings as the thermal-magnetic circuit breakers (of the corresponding group). Circuit interrupters of FA, FH, KA, FH, LA, LH, MA, NH, PA, PH, and PC types are available in the Square D line, including both standard and I-75,000 interrupting ratings.

3-6   CIRCUIT BREAKER HANDLES, LOCKS, AND TERMINAL LUGS

The following descriptions apply primarily to the commercial and industrial circuit breakers in the Square D line.

3-6.1   *Circuit Breaker Handles*

All Square D circuit breakers are provided with a positive quick-make, quick-break, overcenter, trip-free toggle mechanism, shown in Fig. 3–9. All multipole breakers have a single operating handle which acts directly against the contact arms by means of a solid crossbar, assuring positive action on manual and automatic switching. When the circuit breaker is automatically or manually tripped, the handle assumes a center position between ON and OFF. This contrasts with the handle position during other operations. *Center-position trip indication* is a feature of all Square D circuit breakers. In addition, the Q0, Q1, EH, and some Q2 breakers have a *red indicator* that appears in a window of the breaker to indicate the tripped position. VISI-TRIP® is the name given to this added trip indication feature. VISI-TRIP permits easy identification of a tripped circuit, even by people who may have no knowledge of electrical equipment.

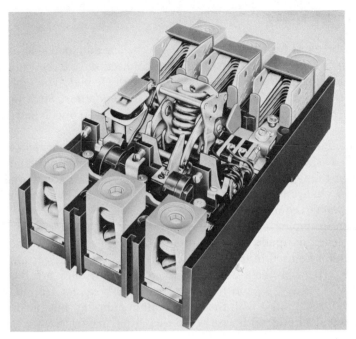

FIGURE 3-9 Circuit breaker toggle mechanism. (Courtesy Square D Company.)

3-6.2 *Padlock and Lock-off Attachments*

All Square D molded-case circuit breakers have attachments that permit the locking or padlocking of the breaker handle in the OFF or ON positions. These attachments are used for safety reasons and/or to prevent routine switching of such circuits as clocks, emergency lights, and so on. The attachments do not interfere with automatic tripping under overload or short-circuit conditions.

A new concept in padlock and handle lock-off attachments is available on Square D circuit breakers of Q2, FY, FA, KA, LA, MA, Q4, and NA types. The attachment is a device that permits the breaker to be locked OFF or ON, or padlocked OFF or ON, any time the need arises. Molded holes in the breaker covers are provided to mount these attachments.

Separate handle lock-off and padlock attachments are available for the Q0 and Ql circuit breakers. These attachments are for field installation. The two- and three-pole QOU circuit breakers have a padlock attachment permanently built into the breaker during manufacturing as a standard feature.

3-6.3   *Terminal Lug*

All circuit breakers have provisions on both the line and load sides for making connection in the electrical circuit. These connections can be plug-on, bolt-on, solderless lugs, wire binder screws, or crimp-type lugs. The most common type of terminal lugs found in industry are the *solderless*. Such lugs are UL listed for the application and are approved for both copper and aluminum cable.

Most industrial circuit breakers are provided with *front-removable lugs,* such as those shown in Fig. 3-10. Occasionally, a lug must be changed due to a change in the type of connection to the breaker, or for some other reason. When a lug change is necessary, being able to change or remove the lug from the front makes the job much quicker and easier.

As an example, lugs that are approved for both copper and aluminum cable are generally made of aluminum. Many industrial users find that aluminum cannot be used in their electrical systems because of certain corrosive conditions. To meet these applications, *copper-only lugs* are available for Square D FA, KA, LA, and MA breakers. The breakers are manufactured and stocked as standard with the copper–aluminum lug, but the copper-only lugs can be substituted with ease when they are front-removable.

FIGURE 3-10   Industrial circuit breakers with front-removable lugs. (Courtesy Square D Company.)

3-7 AUXILIARY DEVICES FOR CIRCUIT BREAKERS

By adding auxiliary devices, the basic circuit breaker functions can be expanded to include such operations as sensing voltage, alarming personnel of faulty conditions, and disconnecting or controlling associated equipments which are not part of the system. The following paragraphs summarize auxiliary devices that are particularly important to industrial circuit breakers. Figure 3–11 illustrates these auxiliary devices.

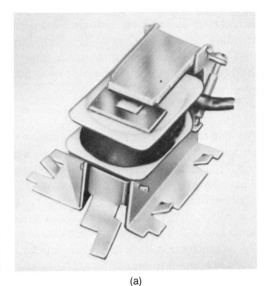

FIGURE 3-11 Typical auxiliary devices for industrial circuit breakers. (Courtesy Square D Company.)

(a)

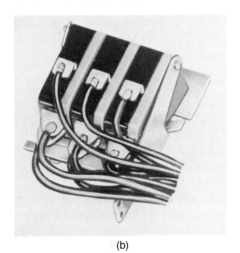

(b)

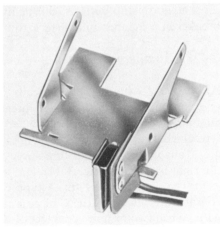

(c)

3-7.1   Shunt Trip for Remote Control

A shunt trip is a mechanism that trips a circuit breaker by means of a trip coil. The coil is energized from a separate circuit or source of power (not part of the circuit being controlled by the breaker), thus providing remote control of the circuit breaker and system. The trip-coil circuit is closed by a relay, switch, or other means. The shunt trip is available in two- and three-pole breakers, and is normally installed in the left pole, as shown in Fig. 3–11.

Shunt trip coils do not have a continuous current rating. A cutoff switch is included to break the coil circuit when the breaker opens. Standard shunt trips are rated 6, 12, 24, 48, 125, and 250 V dc, and 120, 208, 240, 277, 480, and 600 V ac. Other ratings are available upon special request. The control leads for the shunt trip are *color-coded black.*

3-7.2   Undervoltage Trip

An undervoltage trip device trips a circuit breaker automatically when the main circuit voltage decreases to approximately 40% of the normal value. The breaker cannot be reset until the voltage returns to 80% of normal value. These undervoltage trips are available in two- and three-pole breakers, and are installed in the left pole (Fig. 3–11). The trips are supplied as standard in the same voltage ratings as the shunt trips. When the trips are rated above 24 V dc and 240 V ac, external resistors are supplied.

The undervoltage trip is not supplied with time delay action. The control wires for undervoltage trips are *color-coded black.* Since the shunt trip and undervoltage trip are normally furnished in the same pole of the circuit breaker, only the undervoltage trip is necessary to operate both devices when they are installed on the *same* electrical circuit. A normally closed contact, such as those used on stop buttons, can be installed in the control circuit to open the breaker in a manner similar to a shunt trip.

3-7.3   Auxiliary Switches

An auxiliary switch is mechanically operated by the main switching device, and can be used for signaling, interlocking, or other purposes. An A-type auxiliary switch contact is open when the breaker contacts are open, while B-type contacts are closed when the breaker contacts are open. Auxiliary switches are available in two- and three-pole breakers, and are normally installed in the right-hand pole (Fig. 3–11).

The control leads for A-type contacts are *color-coded yellow,* B-type contacts are *blue,* and the C or common lead is coded *blue with yellow tracer wire.* When two or more contacts of the same type are required (two normally open NO, or normally closed NC), the color coding remains the same as described, but the leads are identified by numbered tabs.

3-7.4   Alarm Switch

The alarm switch shown in Fig. 3–11 is a single-pole device activated when the breaker is in the tripped position, and is used to actuate bell alarms and warning signals or lamps. The alarm switch is factory installed in the circuit breaker, and is rated as at least 1 A at 120 V ac.

3-8   SPECIAL APPLICATION CIRCUIT BREAKERS

There are many special application circuit breakers. Two of the most common special breakers used in industry are those with visible blades and those with rear-connecting studs, both of which are shown in Fig. 3–12.

3-8.1   Visible-Blade Breakers

Visible-blade circuit breakers, called Visi-Blade® breakers by Square D, have openings over the contact areas of each pole. The openings are covered with a transparent insulator so that the position of the movable contact is visible. The sides of the openings are painted white, and the tips of the movable contacts are painted red for clear indication. Visi-Blade breakers are available only in standard interrupt capacity type FA, KA, LA, and MA breakers, but not in smaller breakers or the I–75,000 types. Visi-Blade breakers are not UL listed.

3-8.2   Rear-Stud Breakers

Rear-connecting studs are used to assemble molded-case circuit breakers in open air on panel-type switchboards and control stations (Chapter 4). As shown in Fig. 3–12, long and short studs are alternated on adjacent poles to provide adequate voltage spacing between cable connections at each pole. Rear-connecting studs are available for the FA, FH, KA, KH, LA, LH, MA, and MH breakers.

3-9   CIRCUIT BREAKER ENCLOSURES

Most circuit breakers used in commercial and industrial applications are housed in some type of enclosure (panelboard, switchboard, motor control center, individual enclosures, etc.). In this section we concentrate on the individual enclosures. Panelboard, switchboard, and load center enclosures are discussed in Chapter 4. Motor-starter enclosures are covered in Chapter 8.

NEMA (National Electrical Manufacturers Association) has established enclosure designations because individually enclosed circuit breakers are used

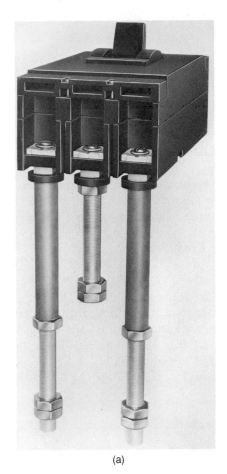

(a)

FIGURE 3-12   Industrial circuit breakers with visible blades and rear-connecting studs. (Courtesy Square D Company.)

in so many different types of locations, weather and water conditions, dust and other contaminating conditions, and so on. The NEMA designation indicates an enclosure type to fulfill requirements for a particular application. Note that the NEMA designations are subject to constant change (as are the requirements of the NEC, local codes, and UL-approved equipment!). The following is a summary of NEMA designations for enclosures used with individually enclosed circuit breakers in commercial and industrial applications.

NEMA 1:   A general-purpose enclosure primarily intended to prevent accidental contact with the enclosed parts. NEMA 1 is suitable for indoor commercial and industrial applications where the enclosure is not exposed to unusual service conditions.

(b)

FIGURE 3-12 *(Cont.)*

NEMA 1A: Now generally obsolete, this was a general-purpose enclosure with gaskets. NEMA 1A has generally been replaced by NEMA 12.

NEMA 3R: A rainproof enclosure intended primarily to protect against a beating rain. NEMA 3R is for all general outdoor applications, except where sleet-proof construction is required.

NEMA 4: A watertight enclosure designed to exclude water applied in the form of a hose stream.

NEMA 5: A dust-tight enclosure. Often, a combination enclosure for both NEMA 4 and NEMA 5 is listed. In most commercial and industrial applications, NEMA 12 can be substituted for NEMA 5.

NEMA 7: An enclosure used where a combustible vapor exists. NEMA 7 meets requirements for NEC, *Class I, Division 1, Group D. (Examples of such an environment are: gasoline vapors, natural gas, paint vapors, and flammable cleaning fluids.)*

NEMA 9: An enclosure used where combustible dust exists. NEMA 9 meets requirements for NEC, *Class II, Division 1, Group E, F, and G. (Examples of this environment are: aluminum dust, coal dust, and flour and grain dust.)*

NEMA 12: An industrial enclosure designed for use in those industries where it is desired to exclude such material as dust, lint, fibers, and flyings, oil seepage, and coolant seepage. Figure 3–13 shows a Square D NEMA 12 circuit breaker enclosure. Such circuit breakers are available with and without knockouts.

3-9.1 Neutrals for Circuit Breaker Enclosures

An insulated neutral (Chapter 5) is provided in all Q1 (100 A) NEMA 1 and NEMA 3 R enclosures used in 240-V systems. Bonding screws and straps are provided, but not installed, where the enclosure may be used as a service entrance (Chapters 1 and 4). Neutrals are available for installation in all other

FIGURE 3-13 NEMA 12 circuit breaker enclosure. (Courtesy Square D Company.)

FIGURE 3-14 Typical circuit breaker enclosure with solid neutral and equipment ground bar. (Courtesy Square D Company.)

Square D enclosures, but are merchandised separately as 100-, 225-, 400-, 600-, 800-, and 1000-A ratings. Figure 3–14 shows a typical circuit breaker enclosure with solid neutral and equipment ground bar.

3-10 ENCLOSED SAFETY SWITCHES

The NEC and common sense requires that all switches be enclosed for safety. Square D Company manufactures two complete lines of enclosed safety switches to meet industrial, commercial, and residential requirements. Both the *heavy-duty* and *general-duty* lines have visible blades (Sec. 3–8.1) and safety handles. With safety handles, the handle is always in control of the switch blades. (If the handle is in the OFF position, the switch is OFF, or open.)

3-10.1 Heavy-Duty Switches

Figure 3–15 shows the construction of a heavy-duty safety switch, together with some typical enclosures. Heavy-duty switches are intended for applications where price is secondary to safety and continued performance. Such switches are usually subjected to frequent operation and rough handling. Heavy-duty switches are also used in atmospheres where general-duty switches are unsuitable, such as automobile manufacturers, breweries, foundries, shipyards, and similar heavy industries.

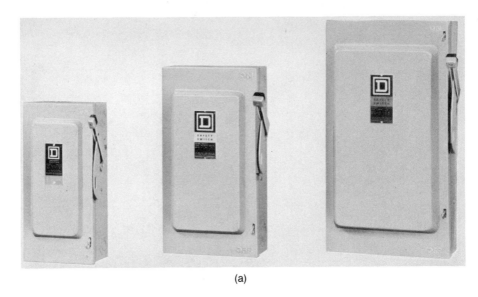

(a)

FIGURE 3-15   Typical heavy-duty safety switch enclosures and internal construction. (Courtesy Square D Company.)

The Square D line of heavy-duty switches are rated at 30 to 1200 A, 240 to 600 V (ac-dc). The switches with horsepower ratings (Chapter 8) are able to interrupt approximately *six times* the full-load, motor-current ratings. When equipped with Class J or Class R fuses (Sec. 3–11), Square D heavy-duty safety switches are UL listed for use on systems with up to 200,000 A available short-circuit current. Heavy-duty switches are available with NEMA 1, 3R, R, 4X, 4, 7, 9, and 12 enclosures.

The heavy-duty switch line of Square D has front-removable, screw-type terminal lugs. All switch lugs are suitable for copper or aluminum wire, except on 30–200 A NEMA 4, 5, 4X, 7, 9, and 12 switches, which are suitable for copper wire only. Switches are UL listed for the wire sizes and number of wires per pole as shown in Fig. 3–16.

3-10.2 General-Duty Switches

Figure 3–17 shows the construction of a general-duty safety switch, together with a typical enclosure. General-duty switches are for light commercial (and residential) applications where the price of the device is a limiting factor. General-duty switches are meant to be used where operation and handling are moderate, and where the available fault current is less than 10,000 A. Some examples of general-duty switch applications are: residential service entrances, light-duty branch-circuit disconnects, major appliance disconnects, and farm or small-business service entrances.

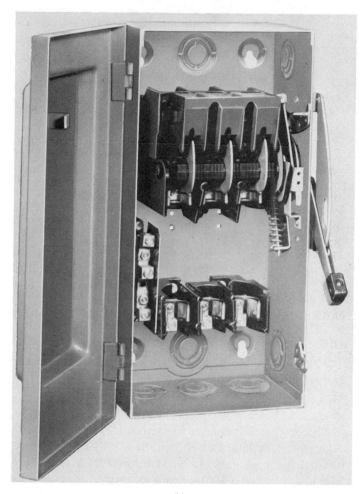

(b)

FIGURE 3-15   *(Cont.)*

General-duty switches are rated up to 600 A at 240 V (ac only) in general-purpose (NEMA 1) and rainproof (NEMA 3R) enclosures. These switches are horsepower rated (Chapter 8) and capable of opening a circuit with *approximately six times the full-load current rating of a motor.*

3-10.3   Double-Throw Safety Switches

Figure 3-18 shows a typical double-throw safety switch. These switches are special-purpose devices used as *transfer switches.* Square D has three lines of double-throw switches: 82,000 line, 92,000 linc, plus a DTU rainproof manual transfer switch.

Ampere rating	Voltage	Conductors per phase	Mechanical lug UL listed wire size
		NEMA 1 and 3R switches	
30	240	1	No. 10–2 Al or No. 14–2 Cu
30	600	1	No. 10–2 Al or No. 14–2 Cu
60	All	1	No. 14–2 Al or Cu
100	All	1	No. 6–1/0 Al or No. 10–1/0 Cu
200	All	1	No. 6–300 MCM Al or Cu
400	All	1	No. 3/0–750 MCM Al or Cu
		2	One no. 3/0–750 MCM Al or Cu
			One no. 6–300 MCM Al or Cu
600	All	2	No. 3/0–500 MCM Al or Cu
800	All	3	No. 4–750 MCM Al or Cu
1200	All	4	No. 4–750 MCM Al or Cu
		NEMA 4 and 5, 4X, 7, 9, and 12 switches	
30		1	No. 14–4 Cu
60		1	No. 14–4 Cu
100		1	No. 14–0 Cu
200		1	No. 6–250 MCM Cu
400		1	No. 3/0–750 MCM Al or Cu
		2	One no. 3/0–750 MCM Al or Cu
			One no. 6–300 MCM Al or Cu
600		2	No. 3/0–500 MCM Al or Cu

FIGURE 3-16 Typical wire sizes and number of wires for heavy-duty switch terminal lugs.

The 82,000 line of switches are available as either fused or unfused devices, and feature quick-make, quick-break action, plated current-carrying parts, a key-controlled interlock mechanism, and screw-type lugs. Arc suppressors are supplied on all switches rated above 250 V. Provisions for up to three padlocks are available to lock the handle in either ON or OFF positions. The 82,000 switches are available in NEMA 1 and NEMA 3R enclosures.

The 92,000 line switches are manually operable, not quick-make, quick-break, and are available as either fused or unfused devices in NEMA 1 enclosures only.

3-10.4   *Summary of Square D Safety Switches*

Figure 3–19 summarizes the safety switches available from the Square D Company.

3-11   FUSE TYPES AND CLASSIFICATIONS

Figure 3–20 shows the current rating and dimensions for a few of the three most common types of fuses: *screw* (or *plug*), *ferrule,* and *knife-blade.* The screw- or plug-type fuse is generally limited to older residential and small of-

FIGURE 3-17 Construction of a general-duty safety switch. (Courtesy Square D Company.)

fice applications. Ferrule and knife-blade fuses are used for the higher currents and voltages found in the commercial and industrial wiring of today.

The following paragraphs describe the classes of UL-listed fuses most commonly used in commercial and industrial wiring.

Class H low-voltage cartridge fuses (also known as NEC fuses) are rated at 600 A or less, 250 or 600 V ac, and have specified dimensions. Class H fuses *are not of the current-limiting type.* UL specifies the interrupting rating of the Class H fuse as 10,000 A (which compares to the 10,000 AIC rating of a standard circuit breaker, described in Sec. 3–2.2). The interrupt rating is not marked on the fuse. Class H is reserved for fuses now tested and listed on the base of the present standard for fuses (UL-198). Note that Class H fuses are not recommended for motor circuits (Chapter 8).

Class J low-voltage cartridge fuses are rated at 600 A or less, 600 V. Class J fuses have an interrupting rating of 200,000 A (which is twice the maximum of any present-day circuit breaker). Class J dimensions are different

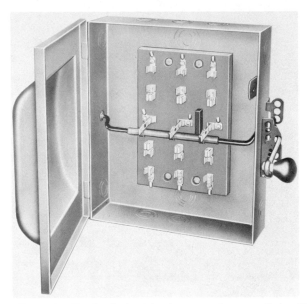

FIGURE 3-18 Typical double-throw safety switch. (Courtesy Square D Company.)

Maximum system voltage	Maximum ampere rating	Switch type	With UL-listed fusing		Type of enclosure
			Class	Short-circuit rating (rms symmetrical amperes)	
General-duty safety switches: single throw					
240 V ac	600 A	Fusible	H	10,000	NEMA 1 and 3R
		Not fusible	—	—	
Heavy-duty safety switches: single throw					
600 V ac and dc	1200 A	Fusible	H	10,000	NEMA 1,
			R	200,000	3R,
			J	200,000	4 and 5,
			L	200,000	4X,
					7 and 9,
		Not fusible			12 and 12 K
Double-throw safety switches					
600 V ac	600 A	Fusible	H	10,000	NEMA 1 and 3R
		Not fusible	—	—	

FIGURE 3-19 Summary of Square D safety switches.

Ferrule			Knife blade		
Amperes	Length	Diameter (inches)	Amperes	Length	Diameter (inches)
30(250 V)	2	9/16	70-100 (250V)	5-7/8	1
30(600 V)	5	13/16	70-100 (600V)	7-7/8	1-1/2
35-60 (250 V)	3	13/16	110-200 (250V)	7-7/8	1-1/2
35-60 (600 V)	5-1/2	1-1/16	110-200 (600V)	9-5/8	1-3/4
			225-400 (250V)	8-5/8	2
			225-400 (600V)	11-5/8	2-1/2
			450-600 (250V)	10-3/8	2-1/2
			450-600 (600V)	13-3/8	3

Screw (plug) type fuses	
Voltage	Amperes
125 V (max)	30 (max)

FIGURE 3-20 Current rating and dimensions of common fuses.

from those of Class H, K, or J fuses. Class J fuses *are current limiting*. The I_P and $I^2 T$ curves (Sec. 3–3) for Class J are less than those of Class K fuses. However, Class J fuses are fast acting, and are selected at 300% of full-load current rating, compared to 125% for Class K time-delay fuses.

Class K fuses are rated at 600 A or less, 250 or 600 V ac. Any of the Class K fuses may be UL listed with interrupting capacities of 50,000, 100,000, or 200,000 A. UL specifies the maximum let-through $I^2 T$ and I_P for each subclass, with K-1 having the lowest let-through (most current limiting) and K-9 (seldom used) with the largest let-through (least current limiting). However, Class K fuses are not allowed to be marked as current limiting, because Class K are physically interchangeable with other types (such as Class H). Class K fuses may also be listed as time-delay if they have a minimum 10-s delay at 500% of rated current. The dimensions of Class K fuses are identical to those of Class H.

Class L low-voltage fuses are rated at 600 to 6000 A, 600 V ac. Class L fuses have an interrupting rating of 200,000 A and have specified dimensions. Class L fuses *are current limiting,* with maximum let-through I_P and $I^2 T$ ratings defined by UL. Class L fuses bolt in place, and do not fit into fuse clips as do smaller fuses.

Class R fuses have the same physical dimensions as Class H or K, but incorporate a notch or ring in one end so that Class R fuses will be accepted in fuse clips designed to reject Class H and Class K. Class RK1 and RK 5 fuses have essentially the same current-limiting effect as Class K1 and K5 fuses, respectively. All Class K1 and K5 fuses with lower let-throughs than the maximum specified for Class R are eligible for relisting as Class R fuses, if the fuses are redesigned by the manufacturer to incorporate the rejection scheme (notch or ring) required for Class R fuses.

4

POWER DISTRIBUTION FROM SERVICE ENTRANCE TO LOAD

In this chapter we discuss distribution of electrical power from the service entrance to the load or electrical outlets. In all electrical systems, residential as well as commercial/industrial, this distribution must include *overcurrent protection*. In all but the simplest systems, the distribution must also include both *feeders* and *branch circuits*. Before we get into the details, let us discuss the basic need for overcurrent protection and feeder/branch circuits.

4-1 OVERCURRENT PROTECTION AND FEEDER/BRANCH-CIRCUIT BASICS

The need for overcurrent protection in all electrical circuits is obvious. Without a circuit breaker or fuse at some point in the circuit, a short in the wiring or in the load (such as an appliance) causes the conductors to overheat. This destroys the conductors and can cause a fire. Typically, circuit breakers and fuses are housed or enclosed in panelboard, switchboards, load centers, and so on. These same enclosures can also house main disconnect devices (mains) and power consumption meters (kilowatthour meters), as discussed in Chapters 1 and 9.

The need for feeders and branch circuits may not be quite so obvious, until you consider the two extremes in alternate design (without feeders and branch circuits).

One extreme is to run a common two-wire system throughout a building and tap off as many outlets and permanent loads as needed. Then provide the common system with overcurrent protection at the service entrance. This overcurrent protection must have an ampacity equal to the total of all the loads. With such an extreme design, individual loads could be destroyed by too much current, without tripping the overcurrent device. On the other hand, if the overcurrent ampacity is lowered to protect the individual loads, the overcurrent device will be actuated by normal loads, when all the loads are used simultaneously.

Another extreme is to use a separate two-wire system for each load, and provide a separate overcurrent device for each two-wire system. Obviously, this is not economical, nor is it practical.

4-1.1   Feeder/Branch-Circuit Relationships

The problems created by either of these extreme design alternatives can be solved with *feeder and branch circuits, which are approved and required by NEC.* The basic feeder and branch-circuit relationship can best be understood by reference to Figures 4–1 and 4–2, which shows the elementary wiring design for a commercial building containing eight separate offices. Note that each office pays its own utility bill and is thus metered separately.

The service entrance (Fig. 4–1) is supplied power through a typical three-wire system (Chapter 1) from the utility company. Thus both 120 V and 240 V are available. One of the three wires is grounded at the service entrance main disconnect (a pull switch in this case) box. (This is the solid neutral wire discussed in Chapter 5.)

From this point, eight feeders supply power to the corresponding eight offices. Each feeder has a separate meter and overcurrent device (a circuit breaker in this case). The meter and circuit breaker are located in an enclosure next to the main disconnect box. However, it is also possible for a service entrance to use panelboards where the mains and feeder circuit breakers are contained within a single enclosure.

Each feeder terminates at a panelboard located within the corresponding office (Fig. 4–2). Note that the three-wire system is carried through to all offices. This is because each office has a separate air conditioning unit, requiring 240 V, in addition to loads requiring 120 V. Each circuit breaker at the service entrance must be two-pole, to accommodate the 240 V.

The office panelboards are provided with eight circuit breakers, one for each branch circuit. Note that the ampacity of all branch circuit breakers is not the same. Also, both one-pole and two-pole circuit breakers are used, to accommodate 120 V and 240 V branches, respectively. Theoretically, an infinite variety of circuit breaker sizes and types can be used in the panelboards. However, there are practical limits set by the availability of standard panelboard sizes and types, as discussed in Sec. 4–2.

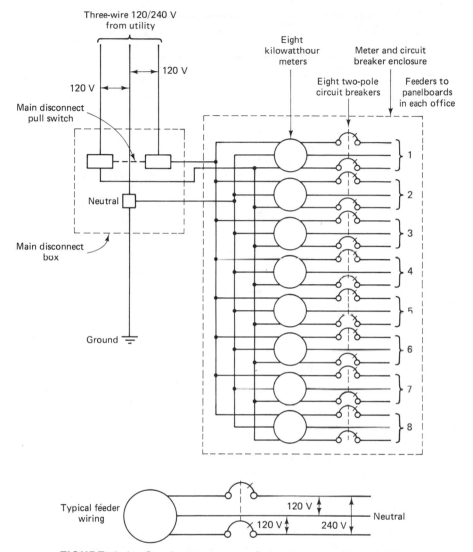

FIGURE 4-1   Service entrance of elementary wiring design for eight-unit office building.

4-1.2   *Feeder/Branch-Circuit Design Steps*

The main steps in design of feeder and branch circuits are as follows:

1. The selection of conductors of the correct size to provide the necessary ampacity.

2. The selection of conductor size to produce a minimum voltage drop. (Although the NEC permits a maximum voltage drop of 5% from

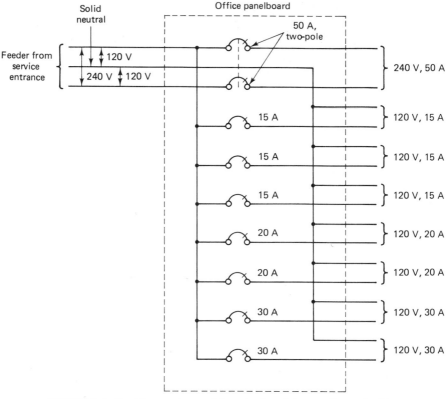

FIGURE 4-2 Elementary wiring design for panelboard of individual offices.

service entrance to final outlet, this percentage is not necessarily efficient.)

3. The selection of overcurrent devices to match the feeder and conductor ampacities.

4. The selection of panelboards, switchboards, load centers, and so on, to accommodate the overcurrent devices and conductors used in the feeders and branch circuits.

5. Finally, and probably most important, the division or arrangement of the loads among the feeders and branch circuits so that there is *no overload* on any one circuit, but each circuit is used to the full capacity. From a cost standpoint, the circuits should be designed so that there is no unnecessary expense caused by an ampacity far beyond that needed for any given circuit.

Each of these problems is discussed in this chapter. We start with a review of the panelboards, switchboards, load centers, and so on, currently available.

4-2 PANELBOARDS AND LOAD CENTERS FOR COMMERCIAL AND INDUSTRIAL USE

This section describes panelboards and load centers manufactured by the Square D Company. Their line is chosen since it represents a cross section of such equipment suitable for commercial and industrial use. Square D manufactures a complete line of general distribution and lighting panelboards, most of which are available either unassembled from distributor stock or factory assembled. One group of panelboards is suitable for use with circuit breakers (Sec. 4-2.1), while the other line is composed of fusable panelboards (Sec. 4-2.2). Square D also offers a line of panelboards known as load centers (Sec. 4-2.3).

4-2.1 Circuit Breaker Panelboards

The four types of circuit breaker panelboards available from Square D are NQO (plug-on) and NQOB (bolt-on) rated for 240 V maximum ac; NEHB (plug-on and bolt-on) rated for 277/480 V maximum ac; and I-LINE, rated for 600 V maximum ac and 250 V maximum dc. All four types are UL listed and meet Federal Specification WP-115a. Also, all Square D panelboards carry UL-listed short-circuit ratings.

NQO Panelboards. The NQO panelboards are rated for use on the following a-c services: 120/240 V single phase, 3 W; 240 V 3 three-phase W delta; 240-V 3 three-phase delta with grounded base phase, and 120/208-V 3 three-phase 4-W wye. The NQO line carries no d-c rating, and is available in assembled or unassembled form. Figure 4-3 shows a typical NQO panelboard suitable for use in industrial buildings, schools, office and commercial buildings, and institutions when the largest branch breaker does not exceed 150 A and the system voltage is not greater than 240 V ac.

NQO panelboards have maximum mains ratings of 400 A, either main breaker or main lugs. Branch circuit breakers may carry the catalog prefix QO, QO-H, QH, Q1, or Q1-H, one-, two-, three-pole, having a maximum rating of 150 A and featuring plug-on bus connections. As discussed in Chapter 3, QO and Q1 breakers are standard with 10,000 AIC rating, and QH breakers have a 65,000 AIC rating. Other ratings for specific applications are also available.

NQO panelboards can also use the Square D QWIK-GARD® branch circuit breakers with ground-fault circuit interruption (GFCI), discussed in Chapter 5. Rated as 10,000 AIC, these GFCI devices provide UL Class A (5 mA sensitivity) ground-fault protection, as well as overload and short-circuit protection for branch-circuit wiring.

NQO panelboards are available in 225-A and 400-A maximum (mains) ratings. The 225-A units are of "single-bus" construction, where one-, two-, and three-pole breakers extend the full width of the panelboard, and cannot

FIGURE 4-3   Typical NQO panelboard. (Courtesy Square D Company.)

be mounted opposite each other. The 400-A units use a "double-row bus" construction which consists of two sets of bus bars mounted on a single pan.

NQO Column-Width Panelboards.   In industrial buildings, the only place to centrally locate lighting panelboards (Chapter 7) is often within the web of a structural steel column. Column-width panelboards, such as those shown in Figs. 4-4 and 4-5, can be used in these industrial situations with a pull box and cable trough (wireway). The neutral connections are made in the pull box to reduce the number of wires in the necessarily narrow wireway. (As discussed in Chapter 2, the number of conductors or wires in a wireway is governed by NEC.)

Column-width panelboards are available with $8\frac{5}{8}$-in.-wide boxes (for use in 10-in.-wide flange structural steel columns) and with $6\frac{7}{8}$-in.-wide boxes (designed for installation in 8-in. WF structural columns). The interiors of column-width panelboards are similar to standard-width interiors, with the exception that the circuit breakers do not twin mount, but are mounted in a single vertical row, as shown in Fig. 4-5. The size of the wiring gutter limits the type of circuit breaker to QO only (Q1 breaker branch conductors require

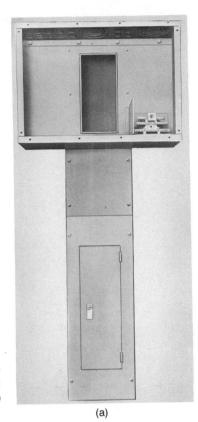

FIGURE 4-4 Column-width panelboard with cable trough and pullbox. (Courtesy Square D Company.)

(a)

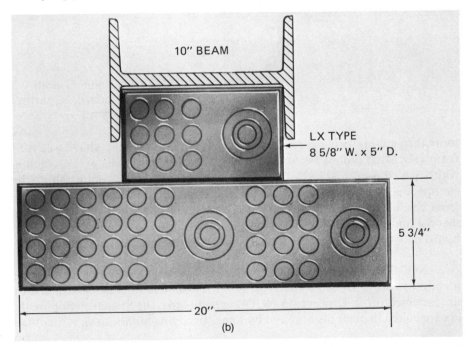

10" BEAM

LX TYPE
8 5/8" W. x 5" D.

5 3/4"

20"

(b)

FIGURE 4-5 Column-width NQO-LX panelboard. (Courtesy Square D Company.)

more than a 2-in. wiring gutter). Due to space limitations, the MONO-FLAT® front (Fig. 4–3) cannot be furnished on either the $6\frac{7}{8}$- or the $8\frac{5}{8}$-in. column-width panelboards. (The MONO-FLAT door hinges and adjustable trim clamps are completely concealed. After the panelboard door is locked, the front cannot be removed, often a desirable feature when used in schools, commercial and industrial buildings, and institutions.) The column-width panelboard door uses conventional hinges and is mounted with external screws.

NQOB Panelboards. The service ratings of NQOB panelboards are identical to those of the NQO type and are available in either assembled or unassembled form. However in NQOB panelboards, the branch circuit breakers are bolted in place (as required by some specifying authorities), rather than plugged into the bus bars. NQOB panelboards are suitable for use in industrial buildings, schools, office and commercial buildings, and institutions when the largest branch breaker does not exceed 100 A and the system voltage is not greater than 240 V ac.

Unassembled NQOB panelboards have a maximum 225-A rating (main lugs or main breaker). Factory-assembled NQOB panelboards are available with maximum 600-A (main lugs) or 400-A (main breaker) ratings. In column-width NQOB panelboards, the circuit breakers have a line-side connection bolted to the main bus.

NEHB Panelboards. As discussed in Chapter 7, many commercial and industrial lighting systems are operated with a 277/480-V three-phase system. The NEHB panelboards are designed to accommodate such lighting systems. Figure 4–6 shows a standard 20-in.-wide NEHB panelboard. Note the main lug/solid neutral compartment (which makes it easier to connect main and branch cables). The NEHB panelboard has the MONO-FLAT front (eliminating any need for interior flush-mounting adjustment) and can accept either plug-on or bolt-on branch circuit breakers.

NEHB panelboards use EH (plug-on) and EHB (bolt-on) circuit breakers with the VISI-TRIP indicator (described in Chapter 3) to provide positive vis-

FIGURE 4-6 NEHB panelboard for commercial/industrial lighting systems. (Courtesy Square D Company.)

ual indication of a tripped breaker. All EH and EHB breakers are UL listed, and all 15-A and 20-A single-pole EH/EHB breakers carry the SWD marking, indicating UL listing as "switching breakers" and attesting to their suitability for switching duty on 277-V or 120-V fluorescent lighting (Chapter 7).

All NEHB panelboards have UL-listed integrated equipment short-circuit ratings of 14,000 A (rms symmetrical) at 277/480 V, and 65,000 A (rms symmetrical) at 240 V, when equipped with the appropriately rated branch breakers. (Chapter 3 describes integrated short-circuit ratings.)

I-LINE Panelboards.   Figure 4–7 shows a distribution panelboard suitable for use with the I-LINE busways described in Sec. 2–8. Figures 4–8 through 4–10 show installation of a circuit breaker within the I-LINE panelboard. With this type of installation, the restriction requiring circuit breakers of the same frame size to be mounted opposite each other (found on some other bus bar panelboards) is eliminated. Each breaker mounts independently and, when required, I-LINE breakers are available with bolted line-side connections.

FIGURE 4–7   I-LINE busway panelboard. (Courtesy Square D Company.)

FIGURE 4-8 Positioning I-LINE circuit breaker on busbar. (Courtesy Square D Company.)

The line connectors are an integral part of the I-LINE breaker, as is the breaker mounting bracket. Only three simple steps are required to add or change an I-LINE circuit breaker:

1. With the panelboard deenergized, the line end of the breaker is positioned on the bus bar stack, as shown in Fig. 4–8. A notch in the base insulator of the bus bar stack fits a ridge at the rear of the breaker

FIGURE 4-9 Levering the I-LINE circuit breaker onto the busbars. (Courtesy Square D Company.)

FIGURE 4-10 Locking the I-LINE circuit breaker in place. (Courtesy Square D Company.)

jaw shroud, for sure "straight-on" seating. The breaker mounting bracket fits neatly into "teardrop" slots in the pan.

2. A screwdriver is inserted in the ratchet slot to lever the breaker firmly onto the bus bars, as shown in Fig. 4-9.

3. The breaker retaining screw is run down, locking the breaker in place, as shown in Fig. 4-10.

Note that MA branch breakers are mounted with a positive cam action rather than being levered in place with a screwdriver. MA breaker mounting requires a $\frac{3}{8}$-in. Allen wrench for installation in an I-LINE panelboard.

I-LINE panelboards are suitable for use on systems rated up to 600 V ac or 250 V dc. Both main-lug and main-breaker panelboards have a maximum rating of 1200 A. Figure 4-7 shows the main lug and solid neutral compartment. With the cover closed, the main lugs are shielded on the top and on each side. The lugs are approved for use with both copper and aluminum cable. When a solid neutral is required, it is mounted adjacent to and in the same compartment as the main lugs. This feature permits all incoming cable to be the same length, thereby saving material, labor, and space. A steel barrier is also provided at the end of the interior opposite the main lugs.

Five types of interior assemblies are used in I-LINE panelboards:

1. HCN interiors are designed to accommodate FY, FA, and Q2 circuit breakers. This allows branch circuits up to 100 A at 240, 277, 480, or 600 V, or branch circuits up to 225 A at 240 V. A maximum main-lugs rating of 600 A and a main-breaker rating of 400 A are available in HCN construction.

2. HCM interiors accommodate any of the circuit breakers described for HCN, plus the KA breaker. This extends the range of branch circuits up to 225 A at 600 V. Maximum main-lugs and main-breaker ratings of 800 A are available in HCM.

3. HCW interiors accept any of the breakers listed for HCN and HCM, plus the LA breaker (one side only). This extends the range of branch ratings through 400 A at 600 V. Maximum main-lugs ratings of 1200 A and main-breaker ratings of 800 A are available in HCW.

4. HCWM interiors accept all the breakers listed for HCN, HCM, and HCW, plus an 800-A MA breaker (when KA, LA, and MA breakers are on one side only). The main-lugs rating is 1200 A. (A maximum main-breaker rating of 800 A is available for HCWM by branch mounting and backfeeding an MA breaker.)

5. HCWA interior is an I-LINE panelboard with 1000-A and 1200-A NH main breakers. Three factory-assembled and one unassembled version of the HCWA interior are available.

The I-LINE panelboards have a NEMA 1 enclosure. A GROUND-CEN-SOR system is preinstalled in the unassembled 1200-A version of the I-LINE panelboard to provide ground-fault protection, as required by the NEC. The I-LINE breakers are provided with the push-to-trip feature (described in Chapter 3), which permits periodic maintenance checks to assure operation of the trip mechanism. Periodic tripping also tests any auxiliary devices associated with the trip mechanism.

Note that I-LINE panelboards have a UL-listed *integrated equipment rating* described in Chapter 3. The integrated equipment rating is the short-circuit rating of the *complete panelboard* with branch circuit breakers installed, and is based on *actual short-circuit tests* with a fault at the load terminals of branch breakers. This is very important in simplified wiring design. For example, if the panelboard has a 100,000-A current interrupt rating (and the appropriate breakers are installed following the manufacturer's recommendations) the system has short-circuit protection of 100,000 A. (From a practical wiring design standpoint, this is sufficient for any application, even heavy industrial wiring systems.)

Also note that the integrated equipment rating also verifies the physical bracing of the bus system and proves the breaker interrupt rating with the breakers installed on the bus assembly. (Prior to integrated ratings, panelboards were component tested and then assembled, but no testing was conducted on the assembly.) In addition to withstanding the primary effects of the short-circuit current (which are the forces trying to blow the panelboard and breakers apart, as discussed in Sec. 2–8.2), the integrated equipment rating also verifies withstanding any secondary effects, such as ionized-gas blow-out.

4-2.2 *Fusible Panelboards*

Square D fusible panelboards (described as QMB) are UL-listed, and meet Federal Specification W-P-115a, Type II, Class 1 for fusible panelboards. Also, fusible distribution panelboards are suitable for use as service equipment, when a main disconnect (main lugs or main breaker) is provided or six or fewer branches are used.

QMB fusible distribution panelboards are rated for use on the following services: single-phase 2 W, single-phase 3 W, three-phase 3 W, three-phase 4 W, and three-phase with grounded base phase, 250 V ac or dc, and 600 V ac. Unassembled QMB panelboards are rated at a maximum 600 A (main lugs only), 200 A (main switch), or 225 A (main circuit breaker). Assembled QMB panelboards are rated at a 1200 A maximum (main lugs) and 800 A (main switch).

As shown in Fig. 4-11, fusible panelboards consist of the fusible switch unit, the interior assembly, and the enclosure (consisting of the box and front). The interior assembly includes the mains, the busing for distributing power to the branches, insulators to support the busing, mounting provisions for the branch protective devices, and a pan to support the entire assembly.

Figure 4-12 shows two fusible switch units. Both units include switches for circuit opening and closing, as well as provisions for mounting the fuses. QMB fusible switch units are available in three forms: (1) switches rated 250 V, ac or dc, 30, 60, and 100 A; (2) switches rated 250 V, ac or dc, 600 V ac, 30-600 A, and (3) fusible interrupters rated 250 V, ac or dc, 600 V ac, 800 A with provisions for Class L fuses. (Fuse classes are discussed in Chapter 3.)

QMB fusible switches are furnished to accept either NEC fuses (also called Class H fuses, as discussed in Chapter 3), Class J fuses, or Class R fuses. Class J fuses are available only on 600-V switches. With the exception of 30-A, two- and three-pole, 240-V general-duty QMB switches, all switches are field convertible to accept only Class R fuses, or can be factory-converted when specified. Unlike Class J fuse provisions, Class R fuse provisions are available on both 250 V, ac or dc, and 600 V ac switches. *QMB fusible switches, with all Class R or all Class J fuses installed, have a short-circuit rating of 200,000 A at 600 V (maximum). This makes fusible switches suitable for the heaviest possible industrial wiring application.*

4-2.3 *Load Centers*

Square D load centers are UL-listed panelboards for use as apartment building distribution equipment (as well as for residential and mobile home use). The Square D load centers shown in Fig. 4-13 use QO, QOT, and Q1 circuit breakers for branch-circuit wiring protection. All Square D load centers meet Federal Specification WP-115A, Type I for use in government housing. Each QO load center has provisions for installation of an equipment ground terminal assembly.

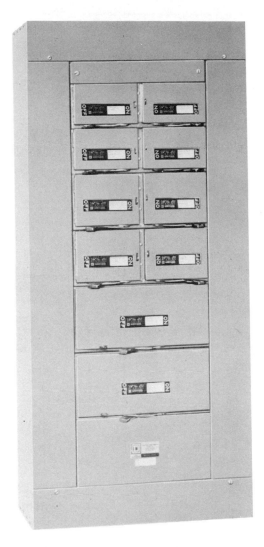

FIGURE 4-11 QMB fusible panelboard. (Courtesy Square D Company.)

The load centers have busing for single-phase 2 W, single-phase 3 W, three- phase 3 W, three-phase 4 W, and three-phase with grounded base phase, at 240 V ac maximum. The load centers are available in three basic types (main lugs, main breaker, and split bus), as well as riser panels.

Main-Lugs Load Centers.   These centers provide distribution of electrical power where a *main disconnect with overcurrent protection is provided separately from the load center.* Single-phase main-lugs load centers are rated 30 through 600 A, while three-phase centers are rated 60 through 600 A. All the terminals are suitable for aluminum or copper conductors. A main lugs load center can be converted to a main-breaker load center by plugging on

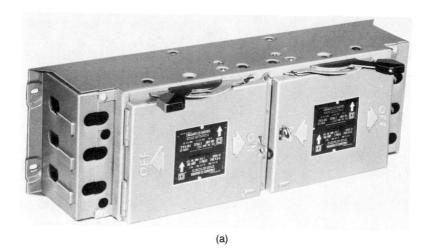

(a)

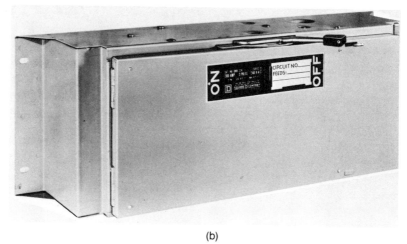

(b)

FIGURE 4-12   Fusible switch units. (Courtesy Square D Company.)

any QO or Q1 circuit breaker, and backfeeding that breaker with the line conductors.

A 600-A, type HQ2 mains-rated load center is available for garden, townhouse, and other types of two- to six-unit apartment complexes where individual metering is not required. The HQ2 is UL listed as *service entrance equipment only,* and the neutral is factory-bonded to the box. The HQ2 center accepts plug-on FA or Q2 two-pole breakers through 200A.

Main-Breaker Load Centers.   Single-phase main-breaker load centers are rated 100 through 600 A, while three-phase centers are rated 125 through 400 A. The main breakers are factory installed. The main breaker and neutral terminals are located at the same end of the load center, allowing straight-in

FIGURE 4-13 Typical load centers. (Courtesy Square D Company.)

wiring and eliminating awkward bends or loops of the incoming service cables. This saves on the total length of the conductors needed. Also, since the neutral location is not in the branch wiring gutter, the load gutters carry only branch breaker connections (and are not cluttered with neutral or ground conductors).

The neutral bars for branch circuits have alternating lugs rated No. 14–8 and No. 14–4 wire size. The bars are UL listed for one conductor per hole and are suitable for copper or aluminum conductors (as are the line-side terminals of the main breakers).

Split-Bus Load Centers. *The NEC allows a maximum of six main disconnects in a common enclosure.* Split-bus QO load centers have the bus split (or divided) into sections which are insulated from each other. This provides an economical service entrance device in those applications not requiring a single main disconnect. The line (or main) section of the load center has provisions for up to six main disconnects for the heavier 240-V loads, subfeeders, and lighting main disconnects. The lower section contains provisions for lighting and 120-V circuits, and is fed by the lighting main disconnect located in the (upper) main bus section. All split-bus devices are provided with factory-installed wires connected to the lower section. These wires can then be field connected to the lighting main disconnect in the main section.

Riser Panels.   Riser panels, consisting of a main-lugs-only load center with an extended gutter of over 6 in., are ideally suited for high-rise office buildings and for apartment complexes. Riser panels are available with 6-, 8-, and 12-circuit load centers.

4-3   CALCULATING FEEDER LOADS

The first step in solving any wiring design problem is to calculate the load or loads. With the load established, and a given source voltage, the required ampacity can be calculated. Then the wire size can be matched to provide the necessary ampacity (with minimum voltage drop), all within the framework of NEC requirements.

A simple way to find the ampacity for a feeder is to add up the maximum ampacity for all branch circuits supplied by the feeder. For example, the ampacity for each feeder in Fig. 4-2 is 195 A (3 × 15 A = 45 A; 2 × 20 A = 40 A; 2 × 30 A = 60 A; 45 + 40 + 60 A + 50 = 195 A). This simple method of calculation could apply to the wiring of Fig. 4-2, but is not adequate for all situations.

The NEC permits the feeder load to be calculated on the basis of the maximum circuit load, in amperes. However, the feeder must have an ampacity only to provide for the maximum *anticipated* load.

4-3.1   Demand-Factor Ratings

The maximum *anticipated* load provision by the NEC is based on the fact that all electrical devices are not used simultaneously, or at their full rating. As an example, Fig. 4-14 shows the demand factors for lighting loads. The NEC includes similar demand-factor ratings for a variety of applications. This means

Occupancy	Portion of lighting load to which demand factor applies (kilowatts)	Demand factor (percentage)
Hotels, motels and apartments (without cooking facilities for tenants)	Up to 20 kW 20 to 100 kW Over 100 kW	50 40 30
All other dwellings	Up to 3.3 kW 2 to 120 kW Over 120 kW	100 35 25
Hospitals	Up to 50 kW Over 50 kW	40 20
Warehouses (storage)	Up to 12.5 kW Over 12.5 kW	100 50
All others	Total	100

FIGURE 4-14   NEC demand factors for lighting and small applicance loads.

that you can use the maximum feeder load to calculate ampacity (by summing all of the branch circuits fed by that feeder), or you can use a lower ampacity based on the NEC demand-factor tables.

As an example, assume that the ampacity of two identical lighting-load feeders is to be calculated, one in a warehouse or storage area, and one in a hospital. In both cases the load is 60 kW. In the case of the warehouse, the lighting feeder ampacity can be calculated on a 50% basis, or as if the load is 30 kW. In the hospital, the lighting feeder ampacity can be calculated on a 20% basis, or as if the load is 12 kW. Of course, the NEC also permits the 60-kW load to be calculated on a full 100% basis, but this results in oversize conductors and overrated breakers.

Figure 4-15 summarizes lighting-load demand factors for various types of buildings, and provides a quick rule of thumb for calculating approximate lighting loads. For example, the anticipated lighting load for 10,000 ft of office space is 50 kW, whereas the lighting load for a bank of the same size is 20 kW.

4-3.2 *Calculating Ampacity for Feeder-Subfeeder Systems*

In the feeder systems described thus far, each panelboard is supplied by a separate feeder, and each feeder is connected directly to the service entrance equipment. Such an arrangement is not satisfactory for large buildings, in-

Occupancy	Load in watts per square foot
Hotels, motels and apartments (without cooking facilities for tenants)	2
Armories and auditoriums	1
Banks	2
Barber shops and beauty parlors	3
Churches	1
Clubs	2
Court rooms	2
Garages (commercial storage)	0.5
Hospitals	2
Industrial commercial buildings	2
Lodge rooms	1.5
Office buildings	5
Restaurants	2
Schools	3
Stores	3
Warehouses (storage)	0.25
Other dwellings (single–family and apartments) 3 kW per square foot	

FIGURE 4-15 NEC lighting load factors.

dustrial plants, and so on, where feeders must run long distances. The heavy currents normally associated with feeders, combined with long runs, produce large voltage drops between the service entrance and panelboards. (The problem of voltage drop is discussed further in Sec. 4–4.) One solution to the problem is to use larger wire sizes for the feeders. This is not economical and, in some cases, is not too practical. For example, if several feeders must be used, the raceway can be overcrowded with large conductors.

A more practical solution is to install a switchboard at some convenient point between the service entrance and the panelboards, as shown in Fig. 4–16. One large feeder supplies the board from the service entrance, and each of the panelboards is served by a subfeeder from the switchboard. The feeder current is the simple sum of the subfeeder currents. The switchboard must provide overcurrent protection for each of the subfeeders. In a practical circuit, where there are many subfeeders, the switchboard can be provided with a main breaker, but that is not required in this case. However, the feeder must be provided with overcurrent protection at the service entrance.

The ampacity of the subfeeder overcurrent devices must be (1) equal to, or greater than, the load ampacity; and (2) equal to, or less than, the ampacity of the corresponding subfeeder conductors. The subfeeder overcurrent devices (probably circuit breakers) must also have an interrupt or withstand rating based on the feeder resistance. For example, if a short or fault occurs just on the load side of any subfeeder breaker, the maximum fault current flows from the service entrance, through the feeder conductors, and through the subfeeder breaker. The current depends on the source voltage and the feeder conductor resistance. There may also be other currents (from other subfeeders), but these are insignificant compared to the fault current.

The ampacity of the feeder overcurrent device must be (1) equal to, or greater than, the combined load ampacities of the subfeeders; and (2) equal to, or less than, the ampacity of the feeder conductors. The feeder circuit

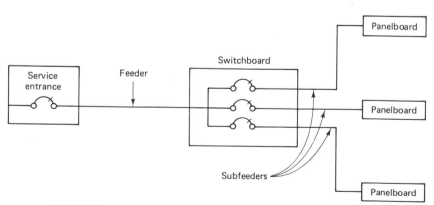

FIGURE 4–16 Wiring with switchboard and subfeeders.

breaker must also have an interrupt rating based on the maximum short-circuit current contributed by the utility company at the service entrance (Sec. 1-3.7).

Since voltage-drop considerations are the primary reason for a feeder–subfeeder system, design examples are given next in Sec. 4-4.

4-4 CALCULATING FEEDER AND SUBFEEDER VOLTAGE DROP

Thus far in this chapter we have discussed feeder ampacity requirements based on anticipated loads. Although feeder conductors must be capable of carrying the full load at all times, feeders and subfeeders must also produce a voltage drop *less than* the permitted limits. *The NEC permits a maximum 3% voltage drop to the farthest outlet in a branch circuit, and a 5% voltage drop if feeders are included.*

This can be interpreted to mean that feeders can have a 2% drop. Of course, if the branch circuits have a 1% drop, the feeders could have a 4% drop, and the system still meets requirements. Except in very unusual circumstances, such an arrangement (4% drop in the feeders) is poor design. Good design dictates that the feeder voltage drop be limited to 2% *maximum*. This allows a full 3% voltage drop for the branch circuits.

4-4.1 Voltage Drop versus Source Voltage

Since voltage drop is calculated as a percentage of the source or service voltage, a larger service voltage permits a larger voltage drop (that is still within tolerance). For example, with a 120-V source and a 2% maximum, the voltage drop can be 2.4 V in the feeder. If 240 V is used, the drop can be increased to 4.8 V, without exceeding the maximum. In practical terms, this means that a smaller wire size can be used for the feeders, with all other factors (distance, load, etc.) remaining equal. Also, a larger service voltage permits a smaller current (for a given load). This, in turn, permits a smaller wire size for the feeders.

A simple solution to the voltage-drop problem is to increase the service voltage. Although simple, the solution is not necessarily practical. In some cases, the service voltage can not be changed. In other cases, the load must be operated from a given voltage. Both of these problems can be overcome by means of transformers. However, the cost and inconvenience of transformers may outweigh the advantages of a higher voltage. In such cases, the designer must decide on a trade-off.

To provide a background for making such a decision, consider the following feeder voltage-drop problem and the four possible solutions: (1) operating at 115/230-V single-phase; (2) operating at 230-V three-phase; (3)

operating at 460-V single-phase; and (4) operating at 115/230-V single-phase, with feeders and subfeeders.

Assume that the load is 30 kW and that the load can be divided into three equal (10-kW) loads, if necessary. The distance from the service entrance to the loads is 500 ft. If necessary, a switchboard can be installed at a convenient point between the service entrance and the load. Circuit breakers are to be used at the service entrance (and at the switchboard, if any). The voltage drop in the feeders (and subfeeders, if used) must not exceed 2%, so there is a 3% voltage drop available for the branch circuits. The load is resistive, and has a power factor of 1. Aluminum THW conductors are to be used throughout. Ignore the effects (derating) of high temperature and skin effect. However, if subfeeders are used, it is necessary to consider the derating effect of having more than three conductors in a raceway.

4-4.2   *Voltage Drop Operating at 115/230-V Single-Phase*

Figure 4-17 shows the basic elements for calculating voltage drop with 115/230-V single-phase service. The first step in any solution is to convert the power into a required current. The 230-V service requires half the current of a 115-V service. Given the larger power involved (30 kW), the 230-V service should be used. Since the load is resistive, the power factor can be ignored (the power factor is 1). With 230 V and 30 kW, the required current is 30 kW/230 V = 131 A. The next standard circuit breaker size is 150 A.

To keep the voltage drop below 2%, select an arbitrary voltage drop below 2%. For example, assume a voltage drop of 1.7% or 230 V × 1.7% = 3.9 V.

Using the information of Sec. 2-4.1, find the wire size for 1000 ft (two times the distance of 500 ft) of aluminum conductors (assumed resistivity of 17).

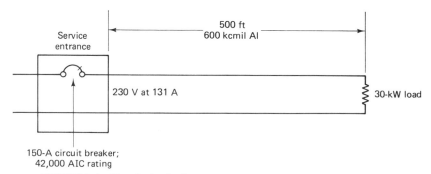

FIGURE 4-17   Calculating voltage drop with 115/230-V single-phase service.

$$\text{cmil} = \frac{17 \times 131 \times (500 \times 2)}{230 \times 0.017} = \text{approx. } 571 \text{ kcmil}$$

The next largest standard-size conductor is 600 kcmil (600 MCM).

Check the selected standard conductor size for correct ampacity. As shown in Fig. 2–7, 600-kcmil aluminum THW conductors can carry more than 131 A (the load) and 150 A (the circuit breaker).

Next, find the d-c resistance of the conductor, and determine the actual voltage drop. As shown in Fig. 2–2, the d-c resistance for 1000 ft of 600-kcmil aluminum conductors is 0.0295 Ω. With this resistance and a current of 131 A, the voltage drop is 0.0295 × 131, or is 3.8645 V (below the desired 3.9 V).

Finally, determine the proper withstand or interrupt current rating for the selected 150-A circuit breaker. Since the circuit breaker is located at the service entrance, the interrupt rating should be based on the utility company's contribution to short-circuit currents (typically, 42,000 A or less, as shown in Fig. 1–9).

4–4.3 *Voltage Drop Operating at 230 V Three-Phase*

Figure 4–18 shows the basic elements for calculating voltage drop with 230-V three- phase service. With the same load operated at three-phase, the load must be divided into three equal parts of 10 kW (to assure a balanced load). However, the total load remains at 30 kW, the loads are resistive, and the power factor can be ignored. With 230 V and 30 kW, the required three-phase current is

$$\frac{30 \text{ kW}}{230 \text{ V} \times 1.732} = 76 \text{ A}$$

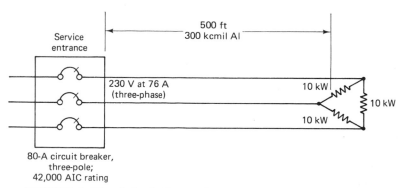

FIGURE 4–18 Calculating voltage drop with 230-V three-phase service.

The next standard circuit breaker size is 80 A.

Again, assume a voltage drop of 1.7%, or 230 V × 1.7% = 3.9 V.

Using the information of Sec. 2–4.1, find the wire size for 1000 ft of aluminum conductors (assumed resistivity of 17).

$$\text{cmil} = \frac{17 \times 76 \times (500 \times 1.732)}{230 \times 0.017} = \text{approx. } 282 \text{ kcmil}$$

The next largest standard-size conductor is 300 kcmil (300 MCM).

Check the selected standard conductor size for correct ampacity. As shown in Fig. 2–7, 300 kcmil aluminum THW conductors can carry more than 76 A (the load) and 80 A (the circuit breaker).

Next, find the d-c resistance of the conductor and determine the actual voltage drop. As shown in Fig. 2–2, the d-c resistance for 1000 ft of 300 kcmil aluminum conductors is 0.059 Ω.

The resistance of each three-phase conductor (500 ft) is 0.059 × 0.5 = 0.0295. With this resistance and a current of 76 A, the voltage drop is 2.242 (well below the desired 3.9 V).

Finally, determine the proper withstand or interrupt current rating for the selected 80-A circuit breaker. Again, the rating is based on the utility company's contribution (typically 42,000 A or less).

From these figures it is obvious that a three-phase 230-V system requires a smaller wire size, and makes it easier to meet a given voltage drop, than does a single-phase 230-V system. However, three-phase is not always available, nor is it possible to operate all equipment with three-phase power.

4-4.4   *Voltage Drop Operating at 460 V Single-Phase*

Figure 4–19 shows the basic elements for calculating voltage drop with 460-V three-phase service. With 460 V and 30 kW, the required current is 30 kW/460 V = 66 A. The next standard circuit breaker size is 70 A. Again, assume a

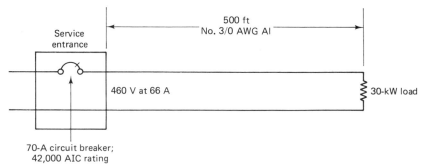

FIGURE 4-19   Calculating voltage drop with 460-V single-phase service.

voltage drop of 1.7%, or 460 V × 1.7% = 7.8 V. Using the information of Sec. 2–4.1, find the wire size for 1000 ft of aluminum conductors (assumed resistivity of 17):

$$\text{cmil} = \frac{17 \times 66 \times (500 \times 2)}{460 \times 0.017} = \text{approx. 144 kcmil}$$

The next largest standard-size conductor is No. 3/0 AWG (Fig. 2–2).

Check the selected standard conductor size for correct ampacity. As shown in Fig. 2–7, No. 3/0 AWG aluminum THW conductors can carry more than 66 A (the load) and 70 A (the circuit breaker).

Next, find the d-c resistance of the conductor and determine the actual voltage drop. As shown in Fig. 2–2, the d-c resistance for 1000 ft of No. 3/0 AWG aluminum THW conductor is 0.105 Ω. With this resistance and a current of 66 A, the voltage drop is 6.93 (below the desired 7.8 V).

Finally, determine the proper withstand or interrupt current rating for the selected 70-A circuit breaker. Again the rating is based on the utility company's contribution (typically 42,000 A or less).

4–4.5   Voltage Drop Operating at 115/230 V Single-Phase, with Subfeeders

Figure 4–20 shows the basic elements for calculating voltage drop with 115–230-V single-phase service, using both feeders and subfeeders. To make the example more practical, the load is considered as three loads at different physical locations, and the loads are not even (one is 12 kW, one 8 kW, and one 10 kW). A switchboard is installed at a common point. The switchboard contains three circuit breakers, one for each load and subfeeder. The switchboard is 250 ft from the service entrance, 100 ft from the 12-kW load, 250 ft from the 8-kW load, and 200 ft from the 10-kW load. The load is resistive, so the power factor can be ignored. The load must be operated from 115/230 V. No more than three conductors are required in any raceway. The maximum 2% voltage drop (230 × 0.02 = 4.6 V) should be divided between the feeders and subfeeders. To simplify calculations, allow a 2-V drop for the feeders and a 2-V drop for the subfeeders.

Design starts with the subfeeders. The required currents for the subfeeders are

$$12 \text{ kW}/230 \text{ V} = 53 \text{ A}$$

$$8 \text{ kW}/230 \text{ V} = 35 \text{ A}$$

$$10 \text{ kW}/230 \text{ V} = 44 \text{ A}$$

The next standard circuit breaker sizes are 60 A, 40 A, and 45 A, respectively. These values can be used. However, to simplify the problem, let

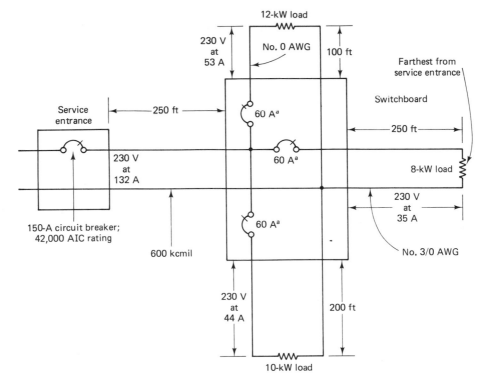

^a60-A circuit breakers in switchboard should have 12,000 AIC rating
(or 10,000 AIC; see text).

FIGURE 4-20 Calculating voltage drop with 115/230-V single-phase, with subfeeders.

us use three 60-A circuit breakers in the switchboard, one for each of the loads
and subfeeders.

The current for the feeder is the simple sum of the subfeeder currents,
or 132 A. The next standard circuit breaker size is 150 A (to be installed at
the service entrance).

The wire size for the 12-kW subfeeder is

$$\text{cmil} = \frac{17 \times 53 \times (100 \times 2)}{2} = \text{approx. 90 kcmil}$$

The next largest standard-size conductor is No. O AWG (Fig. 2–2).

As shown in Fig. 2–7, No. 0 AWG aluminum THW conductors can carry
more than 53 A (the load) and 60 A (the circuit breaker). As shown in Fig.
2–2, the d-c resistance for 1000 ft of No. 0 AWG aluminum conductor is 0.168
Ω. The resistance for 200 (100 × 2) ft is 0.168 × 0.2 = 0.0336 Ω. With this
resistance, and a current of 53 A, the voltage drop is 1.78 V (below 2 V).

The wire size for the 8-kW subfeeder is

$$\text{cmil} = \frac{17 \times 35 \times (250 \times 2)}{2} = \text{approx. } 149 \text{ kcmil}$$

The next largest standard-size conductor is No. 3/ 0 AWG (Fig. 2–2).

As shown in Fig. 2–7, No. 3/0 AWG aluminum THW conductors can carry more than 35 (the load) and 60 A (the circuit breaker). As shown in Fig. 2–2, the d-c resistance for 1000 ft of No. 3/0 AWG aluminum conductor is 0.105 Ω. The resistance for 500 (250 × 2) ft is 0.105 × 0.5 = 0.0525 Ω. With this resistance and a current of 35 A, the voltage drop is 1.84 V (below 2 V).

The wire size for the 10-kW subfeeder is

$$\text{cmil} = \frac{17 \times 44 \times (200 \times 2)}{2} = \text{approx. } 150 \text{ kcmil}$$

The next largest standard-size conductor is No. 3/0 AWG (Fig. 2–2).

As shown in Fig. 2–7, No. 3/0 AWG aluminum THW conductors can carry more than 44 A (the load) and 60 A (the circuit breaker). As shown in Fig. 2–2, the d-c resistance for 1000 ft of No. 3/0 AWG aluminum conductor is 0.105 Ω. The resistance for 400 (200 × 2) ft is 0.105 × 0.4 = 0.042 Ω. With this resistance and a current of 44 A, the voltage drop is 1.85 V (below 2 V).

The wire size for the feeder is

$$\text{cmil} = \frac{17 \times 132 \times (250 \times 2)}{2} = \text{approx. } 561 \text{ kcmil}$$

The next largest standard-size conductor is 600 kcmil (Fig. 2–2).

As shown in Fig. 2–7, 600-kcmil aluminum THW conductors can carry more than 132 (the load) and 150 A (the circuit breaker). As shown in Fig. 2–2, the d-c resistance for 1000 ft of 600-kcmil aluminum conductor is 0.0295. The resistance for 500 (250 × 2) ft is 0.0295 × 0.5 = 0.01475 Ω. With this resistance and a current of 132 A, the voltage drop is 1.95 V (below 2 V).

It will be seen by the figures thus far that *voltage drop* rather than ampacity is the *critical value* in finding wire size. That is, a given wire size may be capable of carrying far more current than the anticipated load, but produces a voltage drop that is very close to the maximum tolerance.

Note that the 8-kW load is farthest from the service entrance (500 ft, 250-ft feeder plus 250-ft subfeeder). The total voltage drop from service entrance to the farthest load is 1.95 (feeder) plus 1.84 (subfeeder), or 3.79 V. This is approximately 1.6% (3.79/230) of the 230-V source. For reference, the voltage drops for the 1–kW and 10-kW loads are 3.73 V and 3.8 V, respectively. Both are well below the maximum 2%.

Finally, determine the proper withstand or interrupt current ratings for the circuit breakers. The rating for the 150-A breaker at the service entrance is based on the utility company's contribution (typically, 42,000 A or less). The ratings for the 60-A breakers at the switchboard are based on the feeder conductor resistance. As discussed, the feeder resistance is 0.01475 Ω. Under a worst-case condition, the maximum short-circuit current flows when a direct short occurs just on the load side of the subfeeder breakers (at the switchboard). Under these conditions, with a voltage of 230 V and a resistance of 0.01475 Ω, the current is approximately 15,333 A. This value can be reduced by 20%, or to 80% of 15,333 A, or approximately 12,000 A. As a practical matter, you can probably get by with standard 10,000-A interrupt circuit breakers for all three of the subfeeders.

4-5   BRANCH-CIRCUIT DESIGN

In practical work, the branch circuits are usually designed first. Then the feeders are designed to match the branch circuits. In some ways, branch circuits are easier to design than feeders. This is because the NEC has a number of specific requirements for branch circuits. In other ways, branch circuits are more difficult to design. The main problem in designing branch circuits is *dividing the load.*

Obviously, any one branch circuit cannot be overloaded. In addition to being poor design, this is a violation of the NEC. On the other hand, it is not economical to have more branch circuits than necessary. It is possible to get this balance in branch-circuit design if several guidelines are followed.

4-5.1   Basic Branch-Circuit Requirements

Figure 4–21 shows basic requirements for branch circuits. As shown, there are five ampacity ratings: 15, 20, 30, 40, and 50 A. These ratings represent the *maximum* to be carried by the branch circuit, as well as the required overcurrent protection for each branch circuit. For example, a 15-A branch circuit requires 15-A overcurrent protection and must be used only where the maximum anticipated current is 15 A. If the maximum current exceeds 15 A, say 18 A, a 20-A branch circuit must be used. If the maximum current exceeds 50 A, at least two branch circuits must be used.

These maximum ratings must be derated to 80% of their ampacity if the load is *continuous*. For example, a 15-A branch circuit is required for a continuous load of 12 A, a 20-A branch circuit is required for continuous loads of 16 A, and so on.

Device	Branch circuit rating				
Minimum size (AWG) for copper conductors:	15 A	20 A	30 A	40 A	50 A
Circuit wires	14	12	10	8	6
Taps	14	14	14	12	12
Fixture wires	18	18	14	12	12
Outlet devices:					
Receptacle rating	15 A	15 or 20 A	30 A	40 or 50 A	50 A
Lampholders	Any type	Any type	Heavy duty	Heavy duty	Heavy duty
Maximum load and overcurrent protection	15 A	20 A	30 A	40 A	50 A
Continuous load	12 A	16 A	24 A	32 A	40 A

FIGURE 4-21   Basic requirements for branch circuits.

The wire sizes shown in Fig. 4–21 are for copper. If aluminum is used, larger wire sizes are required to maintain the ampacity. For example, in a 15-A branch circuit, using aluminum conductors, the circuit wires and taps must be at least No. 12 AWG.

Keep in mind that these wire sizes are *minimum*. It may be necessary to increase the wire size, in special cases, to keep the voltage drop within tolerance. Also, remember that the branch-circuit conductors *must have* an ampacity equal to, or greater than, the overcurrent device. Also, remember that a tap wire can never be larger than a circuit wire. Furthermore, if it is necessary to increase the circuit wire size, the taps and fixture wires must be increased accordingly. Otherwise, the smaller fixture and tap wires do not have adequate protection. Wires used for permanent fixtures can never be larger than a circuit or a tap wire.

The ampacity of outlet receptacles must match the ampacity of the branch circuit. There are two exceptions to this rule. In some cases, 15-A receptacles can be used in 20-A branch circuits; 50-A receptacles can be used in 40-A branch circuits.

Any type of lampholder can be used on 15- and 20-A branch circuits. However, only heavy-duty lampholders can be used on 30-, 40-, and 50-A branch circuits. Heavy-duty lampholders are defined as those with the heavy-duty, or *mogul,* sockets. Such heavy-duty sockets are used on lamps with power ratings over 300 W (300 to 1500 W). Lamps with power ratings below 300 W use medium-base lampholders.

4-5.2   40- and 50-A Branch Circuits

In applications (other than dwellings) 40- and 50-A branch circuits can be used for heavy-duty lampholders, or for any use that requires large currents. However, if medium-base lamps (below 300 W) must be used in any application, they can be tapped into 40- or 50-A branch circuits.

4-5.3   30-A Branch Circuits

In applications (other than dwellings) 30-A branch circuits can be used for heavy-duty lampholders, and for any use that requires large currents (30 A maximum, or 24 A continuous), except lampholders with medium-size sockets.

4-5.4   15- and 20-A Branch Circuits

The 15- and 20-A branch circuits are used extensively for office lighting and small electrical equipment. Any size of lampholder can be used on 15- and 20-A branch circuits, since the overcurrent limit is low enough to protect any one lighting unit. For example, assume that a mogul-base lampholder is tapped to a 15-A branch circuit and that a 1500-W lamp is used with a source of 115 V (all of these are extreme cases). The resultant current flow is about 13 A. Even if the load is left on continuously, the load will not trip the overcurrent.

In theory, there is no maximum number of 15- and 20-A branch circuits that can be used. In practical terms, however, the number of overcurrent devices is limited by available panelboard equipment. Typically, panelboards are limited to 42 overcurrent devices (and usually less if larger circuit breakers are used).

A 15-A branch circuit must have 15-A receptacles. A 20-A branch circuit can have either 15- or 20-A receptacles. This simplifies the use of small portable electrical devices (office typewriters, industrial drills, etc.).

4-5.5   Connecting Electrical Devices to Branch Circuits

If a permanently installed device is connected to a branch circuit and the device is the *only load* on that branch circuit, the current rating of the device cannot exceed 80% of the branch-circuit rating. For example, if a 15-A branch circuit is used exclusively for a permanently installed device, the device can have a current rating up to 12 A.

The matching of receptacles is a good design practice. If a 24-A device is connected to a 15- or 20-A branch circuit, the overcurrent device is actuated. If the 24-A device is connected to a 40- or 50-A branch circuit, a fault in the device can destroy the appliance or cause a fire without opening the higher-rated overcurrent device.

If a permanently installed device is used on a branch circuit with other loads, the current rating of the appliance cannot exceed 50% of the branch circuit. Using the same 15-A branch circuit for example, the device current rating must be 7.5 A or less.

If a portable device is used, the current rating of the device should not exceed 80% of the branch circuit. Most portable devices are designed to ensure this condition. That is, a device with a 15-A plug does not draw more than 12 A. Of course, the designer has no control over the user of portable electric devices. There is no way of preventing two 12-A appliances from being connected at two receptacles of the same 15-A branch circuit. However, under those conditions, the overcurrent device is tripped, and there is no danger to the device or wiring.

4-6 BRANCH-CIRCUIT LOADS AND VOLTAGE DROPS

As discussed, the NEC permits a 3% voltage drop to the farthest outlet in a branch circuit. This drop is measured from the panelboard to the outlet. An alternative method for computing maximum voltage drop is 5% from the farthest outlet to the service entrance. This voltage drop includes the feeder. In theory, if the feeders produce a 1% drop, the branch circuit can have a 4% drop. However, good design dictates that the branch circuit voltage drop (panelboard to farthest outlet) be not more than 3%.

Figure 4–21 shows the requirements for minimum wire size in branch circuits. All branch circuits should be designed with the assumption that the wire sizes given in Fig. 4–21 are used. That is, the length of the branch circuit should be limited so that the voltage drops, using the Fig. 4–21 wire sizes, are less than 3%.

If the branch circuit has only one tap (receptacle, permanently wired fixture, etc.), the voltage-drop calculations are fairly simple (somewhat like those for feeders, Sec. 4–4).

If the branch circuit has several taps, voltage-drop calculations are quite complex. For example, as shown in Fig. 4–22, a 15-A branch circuit has three outlets, each with a 3-A load. The current in the conductors is as follows: 9 A between the panelboard and the first outlet, 6 A between the first and

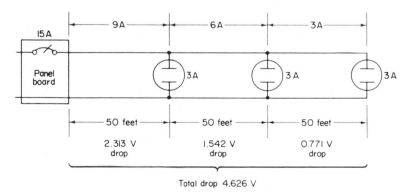

Total drop 4.626 V

FIGURE 4-22 Example of voltage drops in branch circuits.

second outlets, and 3 A between the second and third outlets. Assume that No. 14 AWG copper wire is used (Fig. 4–21), that the resistance is 2.57 Ω per 1000 ft, and that the distances are as shown in Fig. 4–22 (150 ft total).

The voltage drop between the panelboard and first outlet is 100/1000 = 0.1; 0.1 × 2.57 = 0.257 Ω; 0.257 × 9 = 2.313 V.

The voltage drop between the first and second taps is 0.257 × 6 = 1.542. The voltage drop between the second and third taps is 0.257 × 3 = 0.771 V. The total voltage drop is 4.626 (2.313 + 1.542 + 0.771). If the service is 115 V, the maximum permitted voltage drop is 3.45 V. So the circuit of Fig. 4–22 does not meet NEC requirements if the service is 115 V and No. 14 AWG copper wire is used. If the service cannot be raised to 230 V, the wire size must be increased.

This example is oversimplified. The voltage-drop calculations become very complex when the loads, and the distances between taps, are unequal.

4-6.1 Branch-Circuit Guidelines for Voltage Drop

There are some guidelines that can be applied when designing branch circuits. These rules keep the voltage drop within tolerance and still permit use of wire sizes shown in Fig. 4–21. The following is a summary of the guidelines.

1. Keep the number of taps on any branch circuit at six (or fewer) and never more than eight.

2. With a 115-V service, keep the branch-curcuit length to the final outlet at:

   a. Less than 40 ft if the final outlet carries more than 80% of the maximum rated load

   b. Less than 60 ft if the final outlet carries between 50 and 80% of the maximum rated load

   c. Less than 90 ft if the final outlet carrier up to 50% of the maximum rated load

3. If the service voltage is increased, the length may be increased in proportion. For example, if the service voltage is raised to 230 V from 115 V, the lengths may be doubled to 80, 120, and 180 ft.

4. With any service voltage, keep the branch-circuit length from the panelboard to the first outlet at one-third (or less) of the total length of the branch circuit. As an example, assume that the service is 115 V, and that the final outlet carries between 50 and 80% of the total load. Under these conditions, the branch circuit should be less than 60 ft total, from panelboard to the final outlet, and no more than 20 ft from panelboard to the first outlet.

Keep in mind that these are guidelines. The voltage drop must be verified by calculation, especially if there are unusual circumstances (such as long runs between outlets, from panelboard to the first outlet, etc.).

4-7 EXAMPLE OF BRANCH-CIRCUIT, FEEDER, AND OVERCURRENT DESIGN

Assume that an industrial storage area similar to that shown in Fig. 4-23 is to be provided with electrical power. The total area is 15,000 ft² (150 × 100 ft). The storage area is divided into six equal compartments (A, B, C, D, E, and F). Each compartment is 2500 ft² (50 × 50 ft) and must be provided with a separate branch circuit. Each branch circuit has three receptacles, located as shown, and each receptacle must be capable of delivering a *continuous* 5 A of current. A panelboard can be located anywhere in the storage area, but the service entrance must be at one end of the building, as shown. The available service is 115/230 V. Copper TH conductors are to be used throughout. Each branch circuit must be provided with an overcurrent device (circuit breaker) at the panelboard, but the panelboard need not have a main breaker. The feeders must have a circuit breaker at the service entrance.

Find a practical location for the panelboard and routing of the branch circuit raceways. Calculate the required branch circuit ratings (wire size, overcurrent size, and short-circuit rating). Show the raceway routing for the feeders between the service entrance and panelboard. Calculate the feeder wire size, overcurrent size, and short-circuit rating.

4-7.1 Branch-Circuit Ratings

Each branch circuit must be capable of delivering a *continuous* 15 A of current (three receptacles at 5 A each). A 20-A branch circuit rating is thus required in each of the six compartments. (A 20-A branch circuit provides a continuous 16 A, or 80% of 20 A.) Since each of the six branch circuits is 20 A, the panelboard must be provided with six 20-A circuit breakers.

4-7.2 Feeder-Circuit Ratings

The feeders must be capable of carrying a continuous 90 A of current, so a 90-A circuit breaker could be used. However, it is possible that each branch circuit could carry as much as 20 A (say, due to an overload) without tripping the branch circuit breaker. So the feeders should be capable of carrying a maximum of 120 A (6 × 20 A). The next standard circuit breaker size is 125 A.

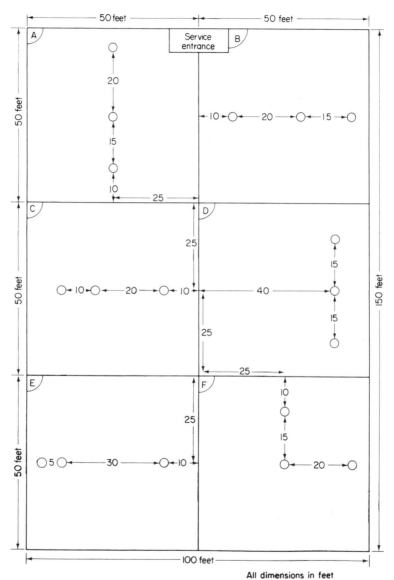

All dimensions in feet

FIGURE 4-23 Physical location of service entrance and loads in storage area.

4-7.3 Branch-Circuit Length

Since all receptacles or outlets provide the same current (5 A), the final outlet carries up to 50% of the maximum rated load. Using the guidelines of Sec. 4-6.1, each branch circuit should not exceed 90 ft if 115 V is used. In our example, we are using 230 V, so the 90-ft length can be doubled to 180 ft for the final outlet. The first outlet must be no more than one-third of the total branch-circuit length, or 60 ft (180/3 = 60).

By placing the panelboard in the center of the building, as shown in Fig. 4-24, it is possible to keep the length of each branch circuit to the farthest receptacle or outlet at less than 95 ft, and the distance to the first outlet at no more than 60 ft.

In compartment A, the length is 25 ft from panelboard to the compartment corner, then 25 ft to the centerline of the compartment, 10 ft to the first receptacle, 15 ft to the second receptacle, and 20 ft to the final receptacle for a total of 95 ft. The same is true of compartments B, E, and F. The distance from the panelboard to the final outlet in compartments C and D is even less (40 and 55 ft, respectively).

Note that all raceways are bent at 90° angles and that no raceway requires a 360° bend (from panelboard to farthest outlet).

4-7.4 Branch-Circuit Wire Sizes

As shown in Fig. 4-21, the circuit wires for a 20-A branch circuit must be No. 12 AWG minimum (when copper is used). Assume that No. 12 AWG copper TH is used from the panelboard directly to each of the receptacles.

4-7.5 Branch-Circuit Voltage Drops

Although the guidelines of Sec. 4-6.1 are calculated so that the branch-circuit voltage drops are less than 3% (from panelboard to farthest outlet), this should be verified by calculating the voltage drop for branch circuit in each of the compartments.

There are several ways to work out this problem. One way is to find the voltage drop between each of the receptacles in the branch circuit, and add the drops. This is very time consuming. The following is a shortcut that can be applied in the majority of cases.

Take the longest branch circuit and assume that the maximum rated (circuit breaker) current flows. If the voltage drop is within tolerance under these conditions, the drop at all other points in any of the branch circuits is well within tolerance.

In our example, the resistance for No. 12 AWG copper wire is 1.62 Ω per 1000 ft. The longest branch circuits (compartments A, B, E, and F) are 95 ft, so 190 ft of conductor is required. The maximum voltage drop is

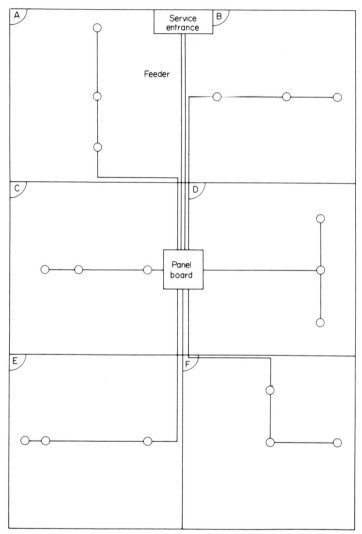

FIGURE 4-24   Feeder and branch circuit routing for storage area.

$190/1000 = 0.190$; $0.190 \times 1.62\ \Omega = 0.3068\ \Omega$; $0.3068\ \Omega \times 20\ A = 6.136$ V. This is below the maximum permitted (3%) drop of 6.9 V (230 V × 0.03).

4-7.6   Feeder Wire Sizes

The feeders run 75 ft from the service entrance (at the end of the building) to the panelboard (at the building's center). The total continuous load is 90 A (15 A for each of the six compartments). However, we are going to base our feeder wire size on a *possible* 120 A of load current and on the 125-A circuit breaker at the service entrance.

Using the information of Sec. 2–4.1, and assuming a voltage drop of 1.7% (230 V × 0.017 = 3.9 V), the wire size for 150 ft (two times the distance of 75 ft) of copper conductors (with assumed resistivity of 10.4) is

$$\text{cmil} = \frac{10.4 \times 120 \times (75 \times 2)}{230 \times 0.017} = \text{approx. } 47.6 \text{ kcmil}$$

The next largest standard-size conductor is No. 3 AWG (Fig. 2–2).

As shown in Fig. 2–7, No. 3 AWG copper TH conductors *cannot* carry the *possible* maximum load. However, No. 1 AWG copper TH conductors can carry more than 120 A (the maximum possible load) and 125 A (the circuit breaker). As shown in Fig. 2–2, the d-c resistance for 1000 ft of No. 1 AWG copper conductor is 0.129 Ω. The resistance for 150 ft of No. 1 AWG is 0.15 × 0.129 Ω = 0.01935 Ω. With this resistance, and a maximum current of 120 A, the voltage drop is 2.322 V (below the maximum of 3.9 V).

4-7.7 Total Voltage Drop

With a feeder voltage drop of 2.322 V, and a branch circuit voltage drop (to the farthest outlet) of 6.136, the total voltage drop is 8.458 V. This is well below the maximum of 5% (230 V × 0.05 = 11.5 V).

4-7.8 Withstand or Interrupt Ratings for Circuit Breakers

The rating for the 125-A breaker at the service entrance is based on the utility company's contribution (typically 42,000 A or less). The ratings for the 20-A breakers at the switchboard are based on the feeder conductor resistance. The feeder resistance is 0.01935 Ω. Under worst-case condition, the maximum short-circuit current flows when a direct short occurs just on the load side of the branch circuit breakers (at the panelboard). Under these conditions, with a voltage of 230 V and a resistance of 0.01935 Ω, the current is approximately 11,900 A. This value can be reduced by 20% or to 80% of 11,900 A, or approximately 9520 A. So standard 10,000-A interrupt circuit breakers can be used for all six branch circuits.

5

GROUNDING COMMERCIAL
AND INDUSTRIAL WIRING

There are two methods for grounding used in commercial and industrial wiring. Figures 5-1 and 5-2 show the basic wiring involved for *grounding* and *grounded* conductors at the service entrance and at downstream panels, respectively. As shown, a grounding conductor is external from the system and is used to protect people from electrical shocks. The grounded conductor is internal to the system and is used to limit available voltage level exposed to people.

Grounding conductors may be green or even bare wires, but are more likely the metal enclosures on raceways and busways housing the circuit conductors. In any case, grounding conductors are used to reduce the possibility of shock by keeping all exposed metallic parts at what can be termed *zero voltage*. Grounding conductors are not considered as current-carrying conductors.

Grounded or neutral conductors are part of the electrical system, and are identified by white or natural gray insulation on the conductor. Grounded wires are connected in various ways to the system in an effort to limit the generated or static voltage level on the system where system parts may be exposed to the operator. Grounded conductors are considered as current-carrying conductors.

Some commercial and industrial wiring requires an undergrounded system. Hospital operating areas are probably the best known ungrounded systems, although there are also some three-phase three-wire ungrounded systems used in industry. The white (grounded or neutral) conductor is not present in

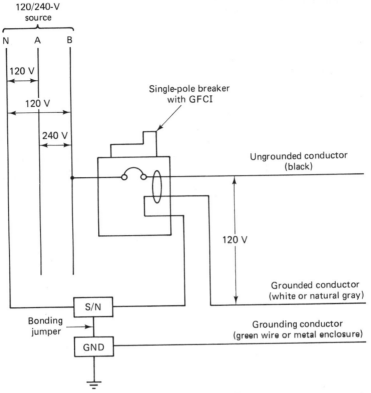

FIGURE 5-1 Basic wiring involved for grounding and grounded conductors at the service entrance.

such systems. However, a green (grounding) conductor, or perhaps a more elaborate grounding system, is present for personnel protection.

5-1 BASIC NEC SYSTEM GROUNDING REQUIREMENTS

As in the case of other wiring standards, the NEC requirements for grounding have changed over the years, and will probably continue to change. However, there are certain basic requirements that both the NEC and common sense dictate. Let us review these requirements before we get into wire sizes and so on (which are subject to change with each new issue of the NEC).

The NEC requires that insulated conductors must be used to carry current to a load and from the load back to the power source. Since it is not practical to insulate the metal raceways and busways used in commercial and industrial systems, all currents must be carried by insulated conductors within

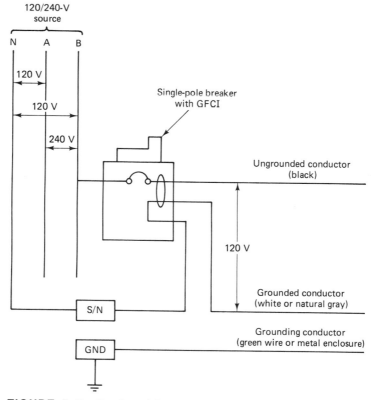

FIGURE 5-2 Basic wiring involved for grounding and grounded conductors at downstream panels.

the raceways. This is in contrast to grounds for electronic devices and automobile wiring, where metal chassis and frames often carry current.

The NEC says that a system must be grounded if, after grounding, the "voltage to ground" is less than 150 V, and recommends that the system be grounded if the voltage to ground can be less than 300 V.

The NEC also states that whatever grounding system is used, the voltage to ground must be the minimum possible voltage. Further, the voltage to ground from an ungrounded point must be the same as the voltage to any grounded conductor.

There are simple methods to meet these basic system grounding requirements.

5-1.1   Voltage-to Ground Problem

A ground can be *true earth ground* or any conductor connected directly or indirectly to earth ground. For example, metal pipes (water, gas, drain, etc.) in a building are connected to other pipes, which, in turn, are buried in the ground. The metal frames of buildings (any conductor touching the metal frames) are installed in the ground. Of couse, some metal frames are sunk in concrete and do not touch true earth ground. However, when concrete is wet or even damp, the moisture can conduct current. So metal pipes and metal building frames can be considered as grounds.

As a practical matter, *the NEC requires that electrical system grounds be made at metal pipes rather than at metal frames,* since buried pipes make for a sure ground. When metal pipes are not readily accessible, a ground for electrical wiring can be made by driving a metal rod into the ground or by burying metal plates in the ground and connecting the wiring to the rod or plate with a suitable conductor.

The *voltage to ground* is the voltage from any point in the electrical system to any object (metal pipe, metal frame, etc.) that is grounded. Let us see how voltage to ground is measured in the three most common commercial and industrial wiring distribution systems.

5-1.2 Single-Phase Three-Wire Ground

Figure 5-3 shows the typical 115/230-V three-wire distribution system used when a single-phase source is used. Note that the *neutral wire is grounded.* If you measure from ungrounded wire to ground, or to a grounded wire, the voltage is 115 V. So the voltage to ground is the minimum possible voltage (per NEC), and the voltage from any ungrounded conductor is the same as the voltage from all other conductors (per NEC).

If either of the other wires is grounded, two NEC requirements are violated. This is shown in Fig. 5-4, where only the bottom wire is grounded.

If you measure from the center or neutral wire to ground, the voltage is 115 V, as shown in Fig. 5-4, which is acceptable. However, if you measure from the top wire to ground, the voltage is 230 V. So the voltage to ground

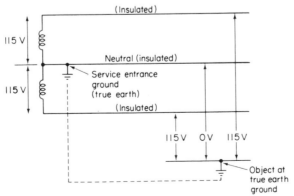

FIGURE 5-3 Single-phase three-wire distribution with correct NEC grounding.

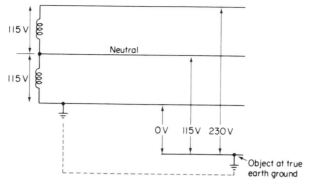

FIGURE 5-4 Single-phase three-wire distribution with incorrect NEC grounding (NEC violation).

from all ungrounded conductors is not the minimum possible, and the voltage to ground from one wire is not the same as from the other wire.

5-1.3 Three-Phase Four-Wire Ground

Figure 5-5 shows the typical 120/208-V four-wire distribution often used for three-phase systems. Again the neutral wire is grounded. In three-phase four-wire systems, *the NEC has an additional requirement that each single-phase load be supplied with one grounded conductor.*

If you measure from any ungrounded wire to ground, the voltage is 115 V. So the voltage is the minimum possible, and the same for all conductors (each load has one grounded conductor, per NEC).

5-1.4 Three-Phase Three-Wire Ground

Technically, an ungrounded three-phase three-wire system does not violate the NEC requirements for grounding. Such a system is shown in Fig. 5-6. If you measure from any wire to ground, the voltage is 0 V. So the voltage is the

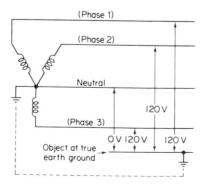

FIGURE 5-5 Three-phase four-wire distribution with correct NEC ground.

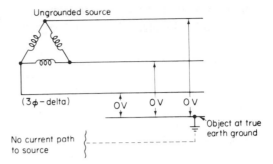

FIGURE 5-6 Three-phase distribution without ground.

minimum possible and the same for all conductors. Although this does meet NEC requirements, and is found in some installations, it is not necessarily the best possible arrangement.

As stated, the NEC *recommends* that grounding be used where the voltage to ground could be less than 300 V. For example, if there is an accidental ground (say because of poor insulation on one wire in a conduit or because of excessive moisture that shorts one wire to ground), the voltage from the other wires to ground is 115 V. This condition is shown in Fig. 5-7.

If a three-phase three-wire system is grounded, *any one of the conductors* can be used for the ground. This meets the NEC requirement that the voltage to ground be minimum and the same for all conductors.

5-1.5 System Ground Considerations

System grounding in commercial/industrial wiring provides a measure of safety should there be a *fault outside the building*. For example, a system ground (where one terminal of the source is grounded) for electrical wiring acts like a lightning rod. This is why *the NEC requires that the system ground be made at the point of service entrance* (where the electrical power enters the building). Should the distribution or service lines be struck by lightning, the service entrance ground provides a ready path to true earth ground *outside* the building, without the danger of heavy current passing through the interior wiring.

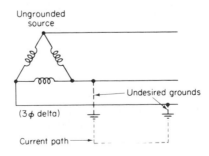

FIGURE 5-7 Three-phase three-wire distribution with undesired grounds (providing a current path between two phases).

5-2   BASIC NEC EQUIPMENT
GROUNDING REQUIREMENTS

Unlike the system grounding discussed in Sec. 5-1, the *NEC requires that all equipment in an electrical system be grounded.* Conduits, raceways, wireways, busways, fittings, boxes, housing, fuse and circuit breaker enclosures, boards, and so on, *must be grounded.* Such grounding is often called *equipment grounding* to distinguish the ground from system grounding. Also, *the NEC requires that all equipment be grounded to the same earth ground as the system (wiring) ground.* The only equipment exempt from grounding is equipment so isolated that no person could touch a grounded object and the electrical equipment simultaneously.

As in the case of system grounding, the main reason for equipment grounding is *safety.* In properly designed electrical wiring, there must be no contact between the electrical equipment (raceways, enclosures, etc.) and the conductors. However, there is always the possibility of an accidental ground, even with a supposedly ungrounded system. The electrical system still operates as if only one of the conductors becomes grounded. However, if a person comes in contact with a true earth ground (say standing on damp cement), and any equipment metal is in contact with a conductor, the person receives the full voltage.

In the example of Fig. 5-8, the conduit has made accidental contact or ground, with one ungrounded conductor. In turn, the conduit is in contact with the switchbox and switchplate (through the plate attaching screws). If a person in contact with true earth ground touches the switchplate, current flows from the fuse, through the conduit, switchbox, switchplate, *and the person,* back to the grounded end of the source. Such a current flow is known as a *fault current,* since the current occurs only when there is a fault in the system.

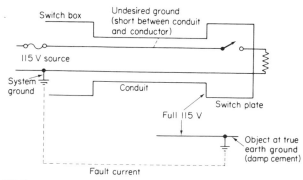

FIGURE 5-8   Effect of short between ungrounded switchplate and conductor, in distribution without equipment ground.

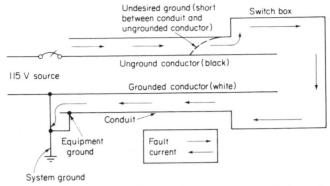

FIGURE 5-9 Effect of short between ungrounded conductors and conduit, in distribution with equipment ground.

Thc purpose of grounding is not to eliminate fault currents but to use the fault currents to open protectivc overcurrent devices (fuses, circuit breakers, etc.).

Of course, grounding all electrical equipment in an electrical system docs not prevent accidental grounding of ungrounded conductors. However, should there by an accidental ground, the overcurrent device opens the circuit to eliminate the hazard.

Such a condition is shown in Fig. 5–9, where the conduit is grounded to thc system ground al the service entrance (as is required by NEC). The circuit breaker is in the ungrounded conductor (also required by NEC). There has been an accidental ground of the ungrounded conductor at the opposite end of the conduit (where the conduit enters the switchbox). Undcr these conditions, current flows through the circuit breaker, opening the circuit.

5-3 COMBINED SYSTEM
AND EQUIPMENT GROUND

The most practical method for grounding both the system and equipment is to provide a common ground point in the service entrance. Usually, this is a terminal bonded to the service entrance enclosure, as shown in Fig. 5–10. In turn, the terminal (sometimes known as the *lay-in-lug*) is connected to a true earth ground (usually a metal pipe) by a conductor of suitable size. As shown, the ground conductor size is determined by the size of the service conductor. For example, if a copper service conductor is No. 2 AWG or smaller, the required size for a copper ground conductor is No. 8 AWG. The sizes shown in Fig. 5–10 must be checked against the NEC of latest issue.

Note that in Fig. 5–10, the neutral system conductor is grounded (as required by NEC) at the service entrance. *The NEC also requires that the neutral conductor not be grounded at any point beyond the service entrance.*

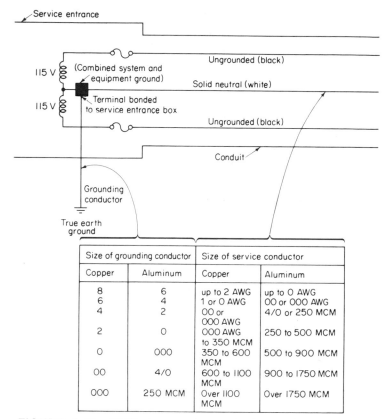

FIGURE 5-10   NEC size requirements for service entrance grounding conductors.

If the neutral conductor is grounded at any other point, there is a possibility that some current could flow through the equipment.

This condition is shown in Fig. 5–11, where the neutral conductor is grounded at the main service entrance and at an interior panel. The service enclosure and the panel are connected by a raceway. Assume that 4 A flows through the load, and assume that the resistance of the equipment (service entrance, raceway, and auxiliary panel) is three times that of the neutral conductor (between the two ground points). Under these conditions, the return current through the neutral conductor is 3 A, with 1 A flowing through the equipment, since the two current paths are in parallel.

5-3.1   *Grounding Interior Equipment*

When it becomes necessary to ground commercial and industrial interior equipment, this should be done with a separate conductor, as shown in Fig. 5–12. The size of equipment grounding conductors is determined by the rating

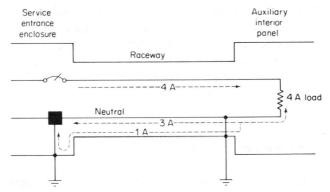

FIGURE 5-11 Effect of grounding neutral conductor at some point in distribution system after the service entrance ground.

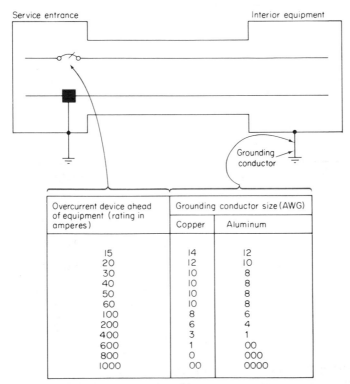

Overcurrent device ahead of equipment (rating in amperes)	Grounding conductor size (AWG)	
	Copper	Aluminum
15	14	12
20	12	10
30	10	8
40	10	8
50	10	8
60	10	8
100	8	6
200	6	4
400	3	1
600	1	00
800	0	000
1000	00	0000

FIGURE 5-12 NEC size requirements for interior equipment grounding conductors.

(in amperes) of the *overcurrent device ahead of the equipment* rather than by the size of the service conductors in the equipment. (Of course, the overcurrent device and conductor sizes are always related, as discussed throughout this book.) As shown in Fig. 5–12, if a 15-A fuse or breaker is used ahead of the equipment, the interior equipment grounding conductor must be No. 14 AWG (for copper) or No. 12 AWG (for aluminum).

5-3.2 Equipment Grounding for Ungrounded Systems

As discussed in Sec. 5–1, all systems are not grounded. Specifically, the three-phase three-wire system does not require a ground. However, *the equipment must be grounded, even in an ungrounded system.* This is shown in Fig. 5–13. With an ungrounded system, the size of the equipment grounding conductor is determined by the size of the service conductor.

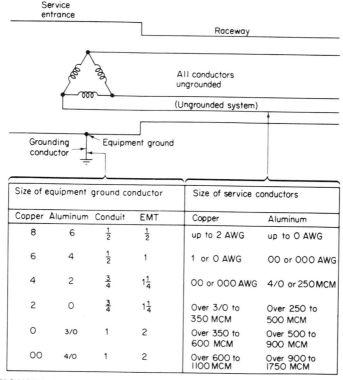

Size of equipment ground conductor				Size of service conductors	
Copper	Aluminum	Conduit	EMT	Copper	Aluminum
8	6	$\frac{1}{2}$	$\frac{1}{2}$	up to 2 AWG	up to O AWG
6	4	$\frac{1}{2}$	1	1 or O AWG	OO or OOO AWG
4	2	$\frac{3}{4}$	$1\frac{1}{4}$	OO or OOO AWG	4/O or 250 MCM
2	O	$\frac{3}{4}$	$1\frac{1}{4}$	Over 3/O to 350 MCM	Over 250 to 500 MCM
O	3/O	1	2	Over 350 to 600 MCM	Over 500 to 900 MCM
OO	4/O	1	2	Over 600 to 1100 MCM	Over 900 to 1750 MCM

FIGURE 5-13 NEC size requirements for equipment grounding conductors for ungrounded distribution systems.

5-4 PROBLEMS OF DEFECTIVE EQUIPMENT GROUNDING

If equipment components are not properly installed, there is a possibility that fault currents (due to an accidental ground) are not sufficient to open over-current devices. Raceways and their associated hardware (connectors and fitting) have low resistance. Most raceways are made of steel or aluminum. These materials have low resistance and, when the raceway components are *properly installed together,* the overall resistance from one end of the equipment remains low. Under these conditions, a fault current opens the overcurrent device and prevents the raceway from being a voltage-to-ground hazard.

If the raceway components, such as clamps, threaded fittings, connectors, bushings, and locknuts, are not making proper electrical contact even though they are mechanically secure, it is possible that the overall resistance of the equipment can be such that fault currents are insufficient to open the overcurrent device. Such a condition is shown in Fig. 5-14. Note that this circuit is similar to that of Fig. 5-10 except that there is some resistance in the raceway. Assume that this resistance is caused by the raceway not making good electrical contact with an interior panel, and that the resistance produced by this poor contact is about 5 Ω.

The condition of poor electrical contact shown here, in itself, does not affect normal operation of the system (since a properly designed electrical sys-

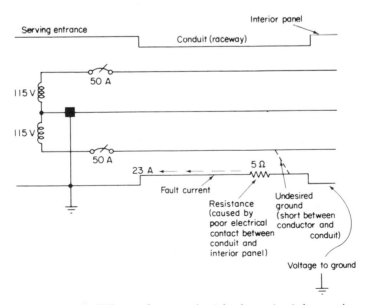

FIGURE 5-14 Effect of poor electrical contact in equipment.

tem does not depend on raceways or equipment to carry normal currents). However, if there is an accidental ground and a current flows, the amount of the current is about 23 A (115/5 = 23).

Assume that the overcurrent device (circuit breaker shown in Fig. 5–14) is rated at 50 A. The 23-A fault current is not sufficient to trip the breaker, and there is some voltage to ground from the interior panel. There are three problems created by this condition.

First, there is the obvious *shock hazard* should anyone touch the interior panel.

Second, if the raceway is making very poor contact with the panel, there can be some *arcing* at the point where the panel and raceway touch. The arcing can produce a fire.

Third, an even greater fire danger exists because of *heating*. Portions of the raceway and interior panel are dissipating 1645 W (115 × 23 = 1645), making the raceway and panel equivalent to a 1600-W electrical heater. Of course, the 5-Ω resistance is not necessarily a practical value and is given here for purposes of illustration. However, the practical problem does exist, and can cause fires. For this reason, some local codes require bonding (by means of a conductor) between raceways and panels or any other place where there is a possibility of poor electrical contact in the equipment.

5-5   PROBLEMS OF DEFECTIVE ARMORED CABLE GROUNDING

In electrical systems that use armored cables as raceways, the cable armor itself provides a path for fault currents. As a general rule, cable armor has less resistance than conduit (designed for the same number of conductors). This is especially true for smaller cables.

The secret of low resistance in armored cable is a ribbon of aluminum that runs the entire length of the cable. The aluminum ribbon is flat and is positioned within the cable between the insulated conductors and the armor. Since the armor and ribbon are in parallel, the overall resistance of the cable from end to end is quite low.

Armored cable must be assembled carefully, particularly where the cable enters outlets, boxes, and so on. Generally, the cable is assembled with clamps to ensure good electrical (as well as mechanical) contact. A problem can develop if the cable is not making good electrical contact. This condition is shown in Fig. 5–15, where a two-wire system is protected by a 20-A breaker. Armored cable is used between the breaker cabinet and the outlet box. The cable is secure at the outlet, but not at the breaker cabinet, and the cable is making intermittent contact with true earth ground (say, a water pipe).

Under the conditions of Fig. 5–15, there is no low-resistance electrical contact between the armored cable or outlet and ground beyond the breaker

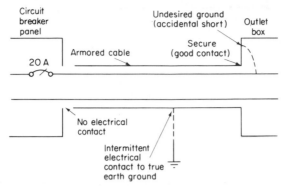

FIGURE 5-15 Effect of poor electrical contact between armored cable and panel, combined with intermittent and accidental contact (undesired grounds).

cabinet. If there is an accidental ground between one conductor and the outlet box, either one of two conditions can occur. If the intermittent contact from the armored cable to ground is one of low resistance and remains long enough, the circuit breaker opens. If the contact is truly intermittent, there is arcing at the contact. In any event, the armored cable and outlet box are "hot" and can be a shock hazard.

5-6 DETECTORS FOR UNGROUNDED SYSTEMS

As discussed in Sec. 5-1, there are no NEC requirements for grounding three-phase three-wire systems. For that reason, ground detector systems are often used with three-phase three-wire. These ground detectors indicate if any one of the three conductors is accidentally grounded. There are a number of schemes used for ground detectors as well as commercial equipment. We will not go into any detail on such systems here. However, one of the simplest ground detectors for 440-V three-phase systems is shown in Fig. 5-16. The detector consists of twelve 115-V lamps connected in groups (four from each conductor to ground).

With all three conductors ungrounded, all lamps glow with the same brilliance. (Note that this is not full brilliance, since each lamp has about 60 to 65 V.) If any one conductor becomes grounded, the group of lamps connected to that conductor goes out. If one conductor has high resistance to ground, the corresponding lamps become dim.

The ground detector of Fig. 5-16 is often used on ungrounded three-phase systems where there are a number of motors, machines, controllers, or similar equipment which have metal surfaces that are readily accessible. If one conductor of such a system contacts the metal surfaces, a serious shock hazard exists.

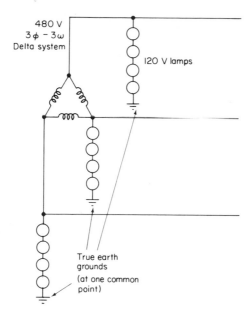

FIGURE 5-16 Simple ground detector for ungrounded three-phase three-wire distribution system.

The detector of Fig. 5–16 can also be used with three-phase four-wire systems (using the appropriate number of lamps). However, such a detector is generally not required, since an overcurrent device opens should one conductor come into contact with a ground. The same is true of single-phase two-wire systems. Of course, as a quick check it is possible to test two-wire systems with a single lamp. Simply connect the lamp from ground to each of the two conductors, in turn. *Only one conductor should cause the lamp to go out.* If the lamp is on for both conductors, one of the conductors is not properly grounded as required by NEC.

5-7   GROUNDS FOR LIGHTING SYSTEM WIRING

As discussed, the NEC requires that one conductor of all two-wire systems be grounded. When this two-wire system is used for lighting, *the NEC further requires that the grounded conductor be connected to the screw shell of the lampholder,* not to the center terminal.

Figure 5–17 shows both the correct and incorrect grounding arrangement for two-wire lighting systems. In Fig. 5–17a (incorrect) the grounded conductor is connected to the center terminal, with the ungrounded conductor at the screw shell. If a person should accidently touch the screw shell and is simultaneously in contact with true earth ground, the person receives the full 115 V, even though not in direct contact with both conductors.

In Fig. 5–17b (correct), the ungrounded conductor is connected to the

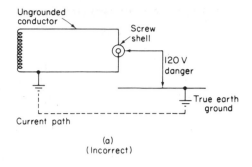

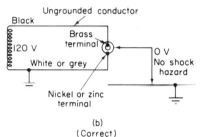

FIGURE 5-17 Correct and incorrect methods of grounding lamp-holders in lighting system design.

center terminal, as required by NEC. There is no voltage to ground from the screw shell, and thus no shock hazard.

5-7.1 Identifying Ground Conductors and Terminals

The NEC requires that the grounded conductor be identified by a white (or neutral gray) outer covering or insulation. Sometimes the words "ground" or "neutral" are stamped on the outer cover of the grounded conductor. If the conductor is larger than No. 6 AWG, the identification (white color) need not cover the entire length of the conductor, but must be at *all terminal ends.* This is usually done with white tape or paint. Any other color, *except green,* can be used for the ungrounded conductors.

The center terminal of lampholders, and the "hot" or ungrounded terminals of receptacles, are identified by brass-colored screws (although the material is not necessarily brass). The screw-shell terminal and grounded receptacle terminal are silver-, nickel-, or zinc-colored. This is shown in Fig. 5–17. Some lampholders and lighting fixtures have wire leads instead of terminals. In these cases, a white (or neutral gray) wire is used to identify the screw shell.

The proper grounding of typical office lamps brings up an obvious problem. Many lamps (particularly older lamps) have two-prong plugs that can be reversed in the receptacle without affecting operation of the lamp. There is no way to assure that the screw shell of the lamp is connected to the grounded conductor. This is the advantage of a three-prong plug, as discussed in Sec. 5–8.

5-8 GROUNDS FOR OFFICE APPLIANCES AND PORTABLE TOOLS

Just as it is possible for a person to touch the screw shell of a lampholder, it is almost impossible to operate office appliances (typewriters, calculators, computers, etc.) or portable power tools without touching the frames or metal surfaces. Of course, such equipment is designed so that neither the ungrounded or grounded conductors make contact with the metal surfaces. However, a defect (such as worn insulation) can occur where one conductor does touch the metal surfaces. This creates an obvious shock hazard, as shown in Fig. 5–18a.

For some years, this problem has been minimized by NEC requirements that most appliances and portable tools be provided with three-wire cords and three-prong *polarized* plugs, as shown in Fig. 5–18b. The third wire, which must have a green outer cover, provides an equipment ground by connecting the appliance or tool metal surface back to the raceway (through the receptacle box). In turn, the raceway is connected to true earth ground, together with the grounded conductors, at the service entrance. Of course, this does not prevent an accidental ground, but should a ground occur, the overcurrent device opens. And there is no shock hazard since the metal surface is at true earth ground.

As shown in Fig. 5–18b, the three-prong plugs are polarized by means of a U-shaped terminal for the equipment ground (green conductor). These plugs do not fit the old-style two-terminal receptacles. This mismatch problem can be overcome. In some appliances and portable tools, the green equipment ground conductor is provided with a lug-type fastener and is separate from the other two current-carrying conductors (which are fitted with a two-prong plug). The terminal lug on the green conductor is attached to the raceway by means of the screw that holds the receptacle cover plate. This is shown in Fig. 5–18c.

Also, there are *adapters* to solve the mismatch problem. These adapters, such as that shown in Fig. 5–18d, have a three-prong receptacle, a two-prong plug, and a ground wire with a terminal lug. The adapter mounts into the wall receptacle by means of the two-prong plug. The adapter ground wire is connected to ground by means of the cover plate screw. The appliance or tool cord (with a new-style three-prong plug) is then connected to the adapter three-prong receptacle.

Virtually all new commercial and industrial wiring uses three-prong polarized receptacles, so the problem should be minimized in the future. When large appliances (such as large office computers) are operated from 230 V with three conductors, the neutral conductor can be connected to metal surfaces or frames, as shown in Fig. 5–18e.

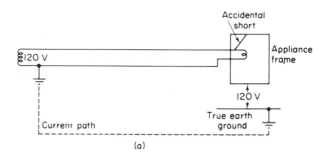

(a)

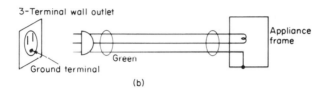

(b)

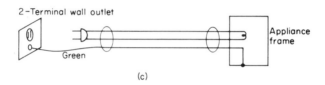

(c)

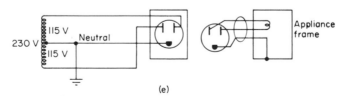

(d)

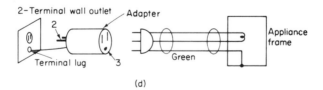

(e)

FIGURE 5-18 Various ground systems for appliances and portable tools.

5-9   SWITCHES AND OVERCURRENT PROTECTION IN GROUNDS

The NEC requires a solid-neutral electrical system. This means that the grounded conductor of any system *must not* be provided with an overcurrent device. Further, *the NEC prohibits using a switch in the grounded conductor,* except where *all* the conductors can be opened by a single switch (two-pole switch for two-wire systems, three-pole switch for three-wire systems, etc.).

Figure 5-19 shows what can happen if both conductors of a two-wire system are provided with fuses (a condition known as *double fusing*). If the fuse in the ground conductor is opened by an overcurrent, the high or "hot" side of the load is at 115 V with respect to ground, and a shock hazard exists. Going further, the ground or low side of the load can also be at some voltage with respect to ground if the load does not drop the entire 115 V. For example, if a person should touch the low side of the load, and the resultant current is sufficient to produce a drop of 65 V across the load, the person receives 50 V.

Figure 5-20 shows incorrect switching and two alternative methods of correct switching for a two-wire system. Figure 5-20a shows what can happen if a two-wire system is provided with a single-pole switch in the grounded conductor (incorrect). When the switch is opened, power is removed from the load. However, the high side of the load is still at 115 V with respect to ground, and a shock hazard exists (as described for double fusing).

In Fig. 5-20b, a single-pole switch is correctly installed in the ungrounded conductor of a two-wire system. Power is removed when the switch is opened. Both the high and low sides of the load are at zero volts with respect to ground.

In Fig. 5-20c, a two-pole switch is installed in the conductors of a two-wire system. Power is removed when the switch is opened and *both sides of the load* are at zero volts. Of course, a two-pole switch is not required for a two-wire system, but this does not violate NEC requirements. As long as all ungrounded conductors are opened simultaneously by a switch or a circuit breaker, there is no shock hazard (and no NEC violation).

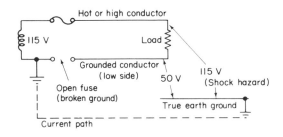

FIGURE 5-19   Effect of double fusing a two-wire distribution system (NEC violation).

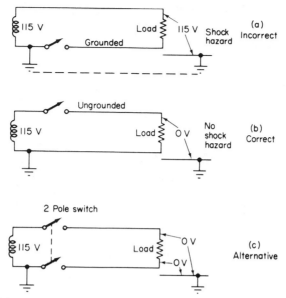

FIGURE 5-20 Correct and incorrect methods of switching in two-wire distribution system.

5-10 GROUND-FAULT CIRCUIT INTERRUPTERS

Ground-fault circuit interrupter (GFCI) devices, shown in Fig. 5–21, function to interrupt the electrical circuit to the load when a fault current to ground exceeds a predetermined value that is less than that required to operate the overcurrent protective device of the supply circuit. (These fault interrupters are also incorrectly known as GFI devices.)

The UL has adopted 5 milliamperes (mA) as the trip level at which devices designated as Class A must operate. The reason for 5 mA is that a current of 5 mA passing through the human body (at a voltage of 120 V) generally causes painful shock. Any current above 5 mA can cause severe injury and possible death. A current of 200 mA or higher results in severe burns, muscle contractions, and probably death by electrocution. However, none of these currents cause a 15-A circuit breaker to open.

GFCI devices are essentially "people protectors." The UL Standard No. 943 defines a Class A GFCI device as one that "will trip when a fault current to ground is 6 mA or more." The tripping time cannot exceed the value obtained by the equation

$$T = \left(\frac{20}{I}\right) 1.43$$

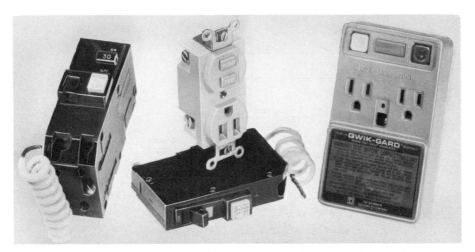

FIGURE 5-21   Typical GFCI devices. (Courtesy Square D Company.)

where T is the time in seconds and I is the ground-fault current in milliamperes. Also, Class A devices must not trip below 4 mA.

Class A GFCI devices must also provide a self-contained means of testing the ground-fault circuitry. To test a Class A GFCI, you simply push the test button. The device then responds with a "trip" indication. UL requires that the current generated by the test circuit not exceed 9 mA, and that the device be functional at 85% of the rated voltage.

The NEC has incorporated several requirements for GFCI units, including requirements for construction sites, outdoor outlets in residential units, boat harbor circuits, receptacles near swimming pools, storable swimming pool circuits, and underwater lighting fixtures in permanently installed pools. Several different types of GFCI equipment have been developed to meet these NEC requirements. We do not cover the entire field of GFCI equipment here, but concentrate on those devices most useful in commercial and industrial wiring systems, including circuit breakers, direct-wire receptacles, and plug-in receptacles.

5-10.1 GFCI Circuit Breakers

Figure 5-22 shows typical circuit breakers that include GFCI protection. These are QWIK-GARD circuit breakers manufactured by Square D Company. QWIK-GARD circuit breakers require the same mounting space as standard QO breakers, and provide the same branch-circuit wiring protection as QO breakers (in addition to the Class A ground-fault protection). QWIK-GARD breakers are UL listed, are available in single- and two-pole construction, with

(a)

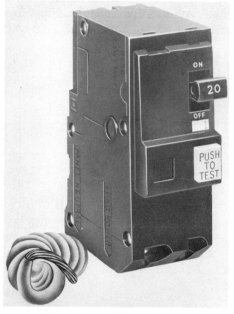

(b)

FIGURE 5-22 Typical GFCI circuit breakers. (Courtesy Square D Company.)

15-, 20-, 25-, and 30-A ratings, and have a 10,000-A interrupt capacity. Single-pole units are rated 120 V; two-pole units are rated 120/240 V. Both single-pole and two-pole units can be mounted in Square D load centers, panelboards, meter pedestals, and Service Pak® power outlets for recreational vehicle parks and construction sites.

Although physically interchangeable with other breakers, GFCI units require special wiring. Both the "hot" and neutral conductors of the circuit to be protected are connected to the breaker, providing the means to pass both the "hot" and neutral current through a *differential current transformer sensor* located within the breaker. Most GFCI units presently manufactured in the United States are designed to operate on the *current imbalance* in the "hot" and neutral conductors that results when a ground fault occurs.

Single-Pole GFCI Sensor Operation.   Figure 5-23 shows how the GFCI portion of a single-pole breaker operates. The GFCI sensor (differential current transformer) continuously monitors the current balance in the ungrounded "hot" load conductor and the neutral load conductor. If a ground fault exists, a portion of the current returns to the source by some means other than the neutral load wire. Under these conditions, the current in the neutral load wire becomes less than the current in the "hot" load wire, and an imbalance between the two load conductors exists. When such an imbalance occurs, the sensor sends a signal to the solid-state circuitry, which, in turn, activates the ground trip solenoid mechanism and breaks the "hot" load connection. A current imbalance as low as 6 mA causes the circuit breaker to interrupt the circuit. This is indicated by the red VISI-TRIP indicator, as well

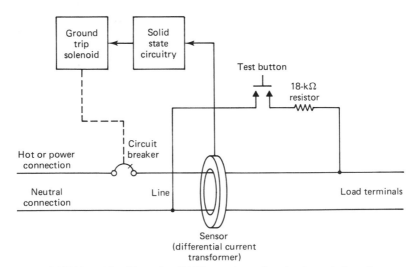

FIGURE 5-23   How the GFCI portion of a single-pole breaker operates.

as the position of the breaker handle (centered between ON and OFF when tripped).

Two-Pole GFCI Sensor Operation.   Figure 5–24 shows how the GFCI portion of a two-pole breaker operates. The GFCI sensor continuously monitors the current balance between the two "hot" conductors and the neutral conductor. As long as the *sum of the three* currents is zero, the breaker does not trip. For example, if there is 10 A current in the A load wire, 5 A in the neutral, and 5 A in the B load wire, the sensor is balanced and does not produce a signal. A current imbalance from a ground-fault condition as low as 6 mA causes the sensor to produce a signal sufficient to trip the breaker.

Testing GFCI Breakers.   The test circuits shown in Figs. 5–23 and 5–24 (a test button and an 18-kΩ resistor) is required by UL, and produces a 6.7-mA simulated ground fault within the breaker when the test button is pressed. (The circuit breaker must be operated at the rated 120 V to produce the full 6.7 mA.) The level of ground fault (amount of fault current) is set by the test circuit resistor (120 V / 18 kΩ = 6.7 mA). When the test button is pressed and the simulated ground fault current flows, the breaker trips just as in an actual ground-fault condition. So operation of the entire circuit breaker is checked. After test, the circuit breaker must be reset in the normal manner.

The UL standards for listing GFCI devices includes tests on operation of the GFCI under many conditions. Specific performance requirements must

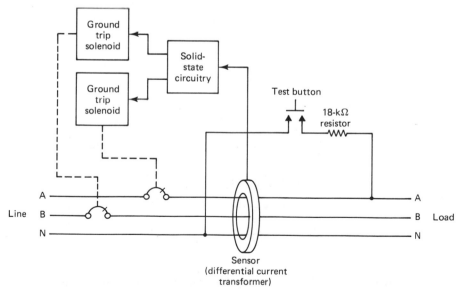

FIGURE 5-24   How the GFCI portion of a two-pole breaker operates.

be met concerning the operation as affected by various conditions, such as voltage changes, ambient temperature, humidity, flooded conduit, and accidental grounding of the neutral conductor of a protected circuit. For example, the typical GFCI circuit breaker must be operated over the temperature range -35 to $+66°C$ $±2°C$, and from 120 to 132 V, for a device rated at 120 V.

Single-Pole GFCI Wiring. Figure 5–25 shows the most common wiring for a single-pole circuit breaker with GFCI. The breaker has two load lugs and a white-wire "pigtail" in addition to the line-side plug-on or bolt-on connector. The line-side "hot" connection is made by installing the breaker in the panel in the same way as any QO or QOB circuit breaker. The white-wire pigtail is attached to the panel neutral (S/N, or solid-neutral) assembly. Both the neutral and "hot" wires of the branch circuit being protected are terminated in the GFCI breaker. These two load lugs are clearly marked LOAD POWER and LOAD NEUTRAL by moldings in the QWIK-GARD breaker case. Also molded in the case is the identifying marking for the pigtail, PANEL NEUTRAL.

Single-pole QWIK-GARD circuit breakers must be installed on *independent circuits*. Circuits that use a neutral common to more than one "hot" conductor *cannot* be protected against ground faults by a single-pole breaker. This is because a common neutral cannot be split and retain the necessary balance between the "hot" wire and neutral wire (under normal use) to prevent the GFCI breaker from tripping. This is discussed further in Sec. 5–10.2.

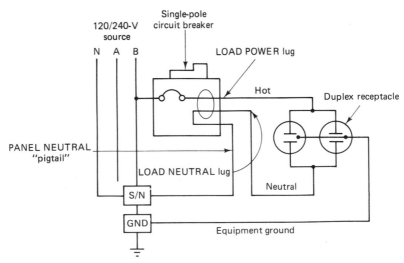

FIGURE 5-25 Typical wiring for a single-pole breaker with GFCI.

Double-Pole GFCI Wiring. Figure 5–26 shows the most common wiring for double-pole circuit breakers with GFCI. A two-pole circuit breaker can be installed on a 120/240-V single-phase three-wire system, on the 120/240-V portion of a 120/240-V three-phase four-wire system, or on two phases and neutral of a 120/208-V three-phase four-wire system. Regardless of the application, the installation of the breaker is the same (connections made to two "hot" buses and the panel neutral assembly). When installed on these systems, protection is provided for two-wire 240-V or 208-V circuit; three-wire 120/240-V or 120/208-V circuits, and 120-V multiwire circuits.

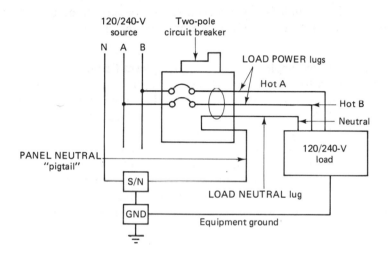

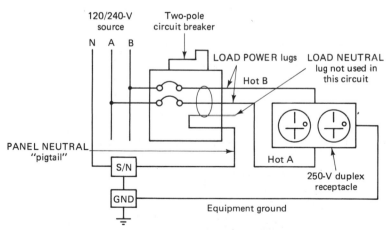

FIGURE 5-26 Typical wiring for a double-pole breaker with GFCI.

5-10.2   *Examples of Ground Faults*
That Trip GFCI Units

It should be noted that GFCI units trip on ground faults, but *provide no protection to the user against contacts that are the same as normal load conditions.* For example, if a person insulated from (or in poor contact with) ground should touch both the "hot" and neutral conductors of a 120-V circuit, this appears as a normal load condition to the GFCI sensor, and the GFCI does not trip. The GFCI trips only if the undesired contact is made between the conductors and ground.

Figures 5–27 through 5–29 show occurrences that appear as ground-fault conditions, and result in the GFCI device opening the circuit.

Ground Faults on Appliances without Equipment Ground.   In Fig. 5–27, a path is established for current flow back to the source by means other than through the branch neutral conductor. Thus the "hot" conductor has less current flowing than the neutral conductor, an imbalance is sensed by the GFCI sensor, and the solid-state circuitry energizes the breaker ground trip solenoid to open the circuit. In this case, the ground fault is caused by an appliance (without equipment ground) connected to the receptacle. It should be noted that GFCI breakers are independent of the equipment ground conductor (Sec. 5–2), and can be used on "older" installations which do not have the equipment ground provisions, as well as on up-to-date installations which do include the ground conductor (Sec. 5–8).

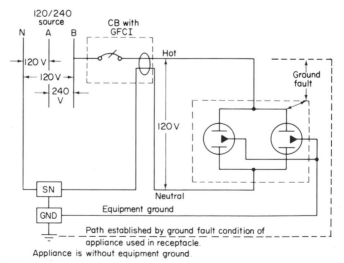

FIGURE 5-27   Effect of ground fault on the sensor within GFCI circuit breaker.

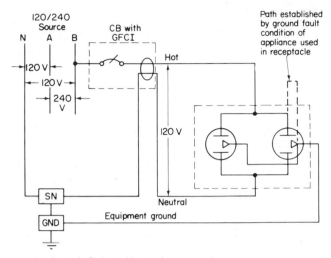

FIGURE 5-28 Effect of faulty appliance (with equipment ground) on GFCI circuit breaker.

Ground Faults on Appliances with Equipment Ground.   In Fig. 5–28, current flows from the source, through the circuit breaker and sensor, to the receptacle. From there, current flows through the faulty appliance (with equipment ground) and back to the source through the equipment ground conductor.

Ground Faults on the Neutral Conductor.   The current path in the circuit of Fig. 5–29 is from the source, through the circuit breaker and sensor, to the receptacle. From there, the current flows through the normal load con-

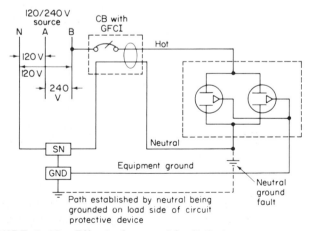

FIGURE 5-29 Effect of ground fault that occurs on neutral conductor at the load side of the circuit breaker.

nected to the receptacle and back to the source through both the neutral and unintended path (established by the neutral being grounded on the load side of the circuit breaker). In such a case, current splits at random between the normal neutral and unintentional path through ground, the neutral conductor carries less current than the "hot" conductor, and the sensor trips the breaker to open the circuit.

5-10.3   *Problem Areas with GFCI Devices*

Proper installation of GFCI units is most important. The following paragraphs describe two of the most common faults in installation of GFCI units.

Transposed Connections.   Figure 5–30 illustrates an *improper installation* that can result from carelessness or from wiring done by an unqualified person. Such a condition can result when adding a subpanel (for branch circuits, not at the service entrance) to an existing system. In such panels the neutral is not grounded, as discussed in Sec. 5-3.

In the improper installation of Fig. 5–30, the subpanel appears only as a 120-V load, the same as other normal loads (lamps, power tools, etc.) that can operate with *either* connection made to the "hot" conductor and the remaining connection made to neutral. However, in the circuit of Fig. 5–30, the GFCI opens under a ground-fault condition *but will not protect the user.* The breaker opens the neutral conductor, leaving the person exposed to the "hot" conductor. *GFCI devices must be installed so as to interrupt all "hot" conductors of the protected circuit.*

Common-Neutral Problems.   Figure 5–31 shows the use of a *common-neutral* for three-wire branch circuits, using breakers *without* GFCI. The same conditions exist if the loads are separate duplex receptacles, or single duplex receptacles that are *split-wired.* (Duplex receptacles can be split-wired when

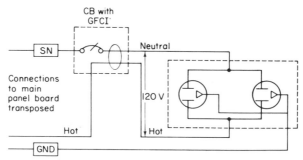

FIGURE 5-30   Improper installation of GFCI circuit breaker on subpanel where connections to main panelboard are transposed.

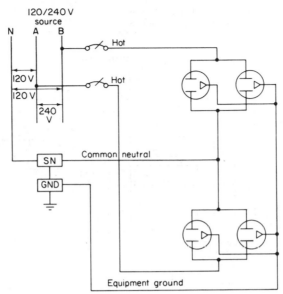

FIGURE 5-31 Typical common-neutral wiring of three-wire system; cannot be used with GFCI circuit breakers.

the duplex arrangement is converted into two single receptacles by removing the interconnecting link between the two "hot" connections on the duplex receptacles.) The common-neutral wiring technique is sometimes used for economic reasons, since some conductors can be eliminated from the branch circuit wiring. However, since single-pole GFCI devices function on the current imbalance between the two conductors passing through the sensor, the *common-neutral wiring technique cannot be used with single-pole GFCI breakers.*

Figure 5-32 shows the use of two single-pole GFCI breakers in a wiring system similar to the common-neutral arrangement. In the circuit of Fig. 5-32, the load neutral connection is not made to circuit breaker 2.

When a load is applied only at receptacle 1, both the "hot" and neutral currents are carried through breaker 1, and no problems exist. However, as soon as a load is applied at receptacle 2, breakers 1 and 2 both trip, since both sensors see unbalanced circuits. Breaker 2 sensor sees only the "hot" current to receptacle 2. Breaker 1 sensor sees the "hot" current to receptacle 1, and the neutral current from both receptacles 1 and 2. Tying the two breaker load neutral terminals together in the panel does not solve the problem, because a load at either receptacle 1 or 2 unbalances both sensors (since the neutral current splits at random through the two sensors).

In some cases neutral conductors for more than one branch circuit are combined under one terminal or connector in a junction box. *This technique cannot be used where GFCI is involved,* since such a connection results in

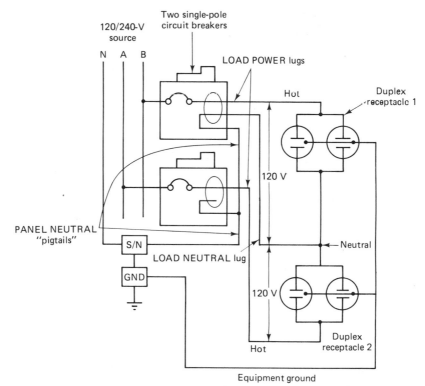

FIGURE 5-32 Improper wiring of two single-pole GFCI circuit breakers with common neutral.

parallel return paths for each of the branch circuits involved (causing an imbalance in the GFCI sensor).

A general rule to remember in GFCI devices is that all conductors (except equipment ground) for a circuit must pass through a single sensor, and these conductors cannot be shared by any other circuit.

Figure 5-33 shows the use of a two-pole GFCI circuit breaker in a common-neutral wiring system. This system eliminates the problems found when trying to use single-pole breakers with a common neutral. Because *both* "hot" currents, and the neutral current, pass through the *same sensor*, under normal load conditions, no imbalance in current occurs among the three elements and the breaker does not trip.

5-10.4 GFCI Direct-Wired Receptacles

Figure 5-34 shows typical direct-wired receptacles that include GFCI protection. These are QWIK-GARD direct-wired receptacles manufactured by Square D Company. The receptacles of Fig. 5-34 provide Class A ground-fault pro-

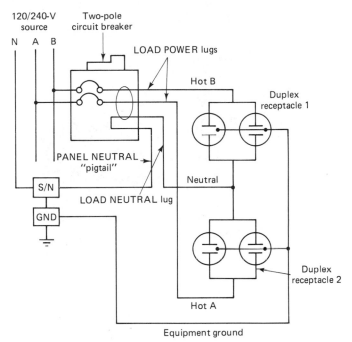

FIGURE 5-33 Two-pole GFCI circuit breaker a common-neutral wiring system.

tection on 120-V a-c circuits, and are available with both 15- and 20-A ratings. The 15-A unit has a NEMA 5-15R receptacle configuration for use with 15-A plugs only. The 20-A device has a NEMA 5-20R receptacle configuration for use with 15- or 20-A plugs. Both 15- and 20-A units have a 120-V, a-c, 20-A *circuit rating.* This is to comply with NEC which requires that 15-A circuits use 15-A rated receptacles, but permits use of either 15- or 20-A rated receptacles on 20-A circuits. Therefore, the 15-A receptacles of Fig. 5–34 may be used on 20-A circuits.

The receptacles of Fig. 5–34 have terminals for the "hot," neutral, and ground wires. Also, the receptacles have feed-through terminals which can be used to provide ground-fault protection for other receptacles downstream on the same branch circuit. All terminals accept No. 14 to No. 10 AWG copper wire. The receptacles have a two-pole tripping mechanism which breaks both the "hot" and neutral load connections. When tripped, the RESET button extends as shown in Fig. 5–35, making visible a red indicating band. The unit is reset by pushing this button.

The receptacles can be mounted, without adapters, in wall outlets that are at least $1\frac{1}{2}$ in. deep. The receptacles have the additional benefit of being *noise suppressed.* Noise suppression minimizes false tripping due to spurious line voltages or radio-frequency (RF) signals between 10 and 500 MHz.

FIGURE 5-34   Typical direct-wired receptacles that include GFCI protection. (Courtesy Square D Company.)

FIGURE 5-35   TEST and RESET buttons on direct-wired GFCI receptacles. (Courtesy Square D Company.)

Keep in mind that GFCI receptacles provide ground-fault protection only, but are not circuit breakers, and do not provide normal branch-circuit protection. However, if GFCI receptacles are used on receptacles, the circuit breakers used to protect the branch circuit need not be GFCI.

5-10.5 GFCI Plug-in Receptacles

Figure 5-36 shows typical plug-in receptacles that include GFCI protection. These are QWIK- GARD plug-in receptacles manufactured by Square D Company. The receptacles of Fig. 5-36 are plug-in ground-fault protection *adapt-*

FIGURE 5-36 Typical plug-in receptacles that include GFCI protection. (Courtesy Square D Company.)

ers for use in either two- or three-wire 120-V a-c receptacles. The plug-in receptacle of Fig. 5–36 has a unique retractable ground pin which makes it possible to provide ground-fault protection at existing two-wire polarized receptacles, as well as on three-wire receptacles. The unit of Fig. 5–36 provides two Class A GFCI-protected receptacles.

To use the plug-in adapter on three-wire receptacles, you lock the ground pin at the back of the unit, as shown in Fig. 5–37. For two-wire receptacles, unlock the ground pin. The ground pin retracts automatically when the unit prongs or stabs are inserted into the receptacle. A yellow indicator pin on the front shows when the ground pin has been retracted. When tripped, the plug-in adapter has a red fault light which illuminates. To reset the unit, simply push the blue reset button.

5–10.6 *Accidental or False Tripping of GFCI Units*

All GFCI devices can be tripped accidentally. Following is a summary of the major causes for such false tripping.

GFCI devices may trip as a result of current leakage to ground. Such leakage can be caused by defective devices or by the accumulative current to many good devices operating on the same circuit (where the leakage is within permissible limits). Leakage can also result from situations such as receptacles becoming wet. In certain cases of current leakage, the circumstances are hazardous, and the GFCI opens the circuit for protection of the user. However,

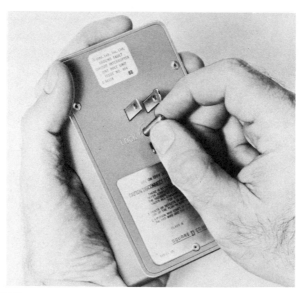

FIGURE 5–37 Locking ground pin on plug-in receptacles with GFCI protection. (Courtesy Square D Company.)

since a GFCI has no way of recognizing circumstances, tripping may occasionally occur when the conditions are not actually hazardous.

When considering possible accumulative current leakages and other causes of GFCI opening of circuits, it is obvious that GFCI protection should be located in *individual* branch circuits, *not in main or feeder circuits*. With a GFCI in a branch circuit, a problem in one circuit does not shut down the entire area or all the power in an office or store. When only the branch circuit is opened, the other circuits remain in operation (and can provide lighting to find the trouble). Also, since GFCI circuit breakers should be tested (by means of the push-to-test provision), the use of GFCI devices in branch circuits results in a minimum of disturbance, since the entire panel is not interrupted.

GFCI devices are sometimes installed on circuits other than those specifically required by NEC. False trippings have been reported on circuits where high-voltage spikes occur during the opening of inductive circuits with relays, contacts, and so on (such as the lighting and motor control circuits described in Chapters 7 and 8). Most problems of this type can be solved by the addition of a capacitor to the inductive circuit being opened. The capacitor must be of proper value to limit the voltage spike to a level the GFCI can withstand. Such problems must be handled on an individual basis.

6

TRANSFORMERS IN COMMERCIAL AND INDUSTRIAL WIRING

For many years transformers have been used extensively in the design of industrial or heavy-duty electrical wiring systems. The main purpose of such transformers is to provide several different voltages, for use by different loads, from a single voltage source. A classic example is where an industrial plant requires 480-V three-phase power for electrical motors, and 120-V single-phase for lighting. Without transformers, it is necessary to supply two separate electrical services to the plant. With transformers it is possible to supply only the 480-V three-phase power to the service entrance and distribute this power to the motors. A transformer at or near the service entrance (or near the load) is connected to convert the 480-V three-phase into 120-V single-phase and distribute the single-phase power to the lighting.

6-1 TRANSFORMER BASICS

6-1.1 Voltage, Current, Impedance, and Turns Ratio

The voltage, current, and impedance ratios of transformers depend on the *turns ratio* of the primary and secondary windings. The calculations for voltage, current, and impedance ratios of transformers are shown in Fig. 6–1. The following are some examples of how these relationships can be used to calculate voltage and currents.

Assume that a transformer must be selected to supply 30 A and 120 V at the secondary winding. The available source is 480 V. Find the primary current using the equations of Fig. 6-1.

The primary current, I_P, is found by

$$I_P = \frac{E_S \times I_S}{E_P} \quad \text{or} \quad \frac{120 \times 30}{480} = 7.5 \text{ A}$$

Assume that another transformer is connected to the same source. The primary has 500 turns, the secondary has 200 turns. What voltage is available at the secondary?

Secondary voltage, E_S, is found by

$$E_S = \frac{E_P \times N_S}{N_P} \quad \text{or} \quad \frac{480 \times 200}{500} = 192 \text{ V}$$

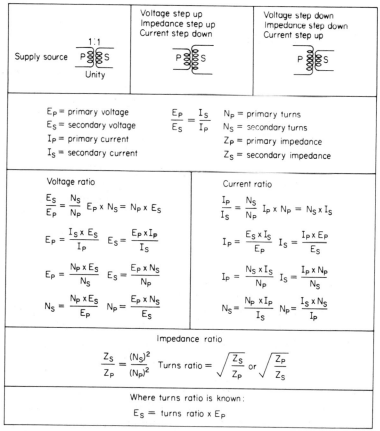

FIGURE 6-1 Calculations for voltage, current, and impedance ratios of transformers.

Assume that the primary of the same 480/192-V transformer is connected to an unknown source. The secondary is connected to an 8-A load. Find the primary current.

$$I_P = \frac{N_S \times I_S}{N_P} \quad \text{or} \quad \frac{200 \times 8}{500} = 3.2 \text{ A}$$

Assume that a transformer shows 240 V on the primary and 80 V at the secondary? What is the impedance ratio (primary to secondary)?

$$\frac{E_S}{E_P} = \frac{N_S}{N_P} \quad \text{and} \quad \frac{E_P}{E_S} = \frac{N_P}{N_S} = \frac{240}{80} = \frac{3}{1}$$

$$\frac{Z_P}{Z_S} = \frac{(N_P)^2}{(N_S)^2} = \frac{(3)^2}{(1)^2} = 9 \text{ (primary-to-secondary turns ratio)}$$

6-1.2 Transformer Winding Polarity Markings

In theoretical problems such as that shown in Fig. 6–1, transformer windings are marked as primary and secondary. Generally, the primary is connected to the source, with the secondary connected to the load. Usually, there is a step up or a step down in turns between the primary and secondary. An *isolation transformer* is an exception to this. Isolation transformers have the same primary and secondary windings, and are used to eliminate any direct contact between primary and secondary circuits (source and load).

In practical applications, transformer windings are generally marked as to high and low voltage rather than primary and secondary. *Low-voltage* windings are typically marked with an X. X1 and X2 are used to identify the opposite terminals of one winding. If there are more low-voltage winding on the same transformer, the secondary winding is marked X3 and X4, the third winding is marked X5 and X6, and so on. *High-voltage* windings are marked with an H (H1 and H2, H3 and H4, etc.).

The numbers used to identify transformer windings are also significant. All even-numbered terminals have the same instantaneous polarity. That is, assuming an alternating current, all even-numbered terminals swing positive at the same time. All odd-numbered terminals swing negative at the same time. All even-numbered terminals are in phase with each other and out of phase with all odd-numbered terminals.

The terminals are not numbered on some small transformers. Instead, the polarity or phase is marked by a dot. All terminals with a dot have the same phase and are out of phase with all terminals that do not have a dot.

Whatever system is used, the polarity markings *must be observed* when connecting transformer windings. For example, assume that a transformer has two identical secondary windings marked H1 and H2, H3 and H4, and each winding produces 120 V. If the odd terminals are connected together (H1-H3, H2-H4), the output is 120 V. However, if an odd terminal is connected to an even terminal (say H1-H4), and the output is taken from the opposite terminals, the output is 240 V (or the sum of the two 120-V windings). The same results are produced when H2 is connected to H3 and the output is taken from H1 and H4.

No matter how transformer windings are connected, *never connect any winding to a voltage higher than the rating for that particular winding.* An exception to this is where two (or more) windings are connected in series. For example, using the same transformer with 120-V windings, if H1 is connected to H4, H2 and H3 can be connected to a 120-V source (provided that the current ratings are not exceeded).

6-1.3 Transformer Power Ratings

In addition to voltage and frequency, transformer windings are rated as to power. Generally, transformers are rated in voltamperes (VA) or kilovoltamperes (kVA) rather than in watts. Since the power factor of the load may be unknown, a rating in watts is difficult to use.

The VA or kVA rating of a transformer shows the safe limit of the load that can be connected (generally to the secondary circuit). For example, if a transformer is rated 3000 VA (or 3 kVA) and the load is connected to a 120-V winding, the maximum safe current in that winding is $3000/120 = 25$ A.

The VA or kVA rating also applies to the source (or primary) winding. For example, assume that the same 3-kVA transformer has a 240-V source winding. Then the maximum safe current in that winding is $3000/240 = 12.5$ A.

When a transformer has more than one coil in either the primary or secondary, the power rating of the transformer must be divided by the number of windings in each section. (In practical applications the winding in each section of most multiwinding transformers have the same voltage rating. This permits the windings to be operated in parallel.)

Assume that the same 3-kVA transformer has two 240-V primary (or source) windings and two 120-V secondary (or load) windings. Then the maximum safe current for the 240-V windings is $3000/2 = 1500$; $1500/240 = 6.25$ A. The maximum safe current for the 120-V windings is $1500/120$ V $= 12.5$ A.

6-1.4 Transformer Losses and Efficiency

All transformers have some losses. That is, all transformers introduce some power loss into the electrical system. Stated another way, the output power from any transformer is always somewhat less than the input power. The lower the output power in relation to input power, the lower the efficiency (as indicated by a lower percentage of efficiency).

There are two types of transformer loss: *copper loss* and *core loss*. Copper loss (also known as I^2R loss, resistance loss, or possibly heat loss) is the result of resistance in the transformer windings. All windings have some resistance that requires power to overcome, resulting in some loss of power. Copper loss varies as the square of the load current. For example, assume a load current of 8 A and a winding resistance of 5 Ω. This produces a copper loss of I^2R, or $8^2 \times 5 = 320$ W (or 320 VA, assuming a power factor of 1). Now assume that the load current is increased to 11 A. This produces a copper loss of $11^2 \times 5 = 605$ W.

Core loss occurs because a small amount of power is necessary to force the flux in the transformer's magnetic core material to alternate at the frequency of the source voltage. Core loss is always present and is independent of transformer loading. However, core loss is calculated on the bases of full-rated voltage. For example, if a transformer is rated at 240 V (primary or source) but is connected to a 120-V source, the core loss is less than if a 240-V source is used.

Copper loss and core loss must be combined to find the total loss of the transformer. In turn, the combined copper and core losses are compared with the transformer's power rating to find efficiency. For example, assume that a 30-kVa transformer has a core loss of 0.3 kVA and a copper loss of 0.4 kVA. The percentage of efficiency is found by

0.3 kVA (core) = 0.4 kVA (copper) = 0.7 kVA (combined loss)

input power = 30 kVA + 0.7 kVA = 30.7 kVA

output power = 30 kVA

efficiency = (output/input) × 100 = (30/30.7) × 100 = 97.7%

Typically, copper losses and core losses are both less than 2% of the transformer's power rating. So the combined losses are less than 4% of the power rating. The efficiency or the combined losses (or both) are sometimes given on the transformer's nameplate. The procedures for measurement of transformer losses are given in Sec. 6–1.6.

6-1.5 Transformer Regulation and Percentage of Impedance

All transformers have some voltage-regulating effect. This is, the output voltage of a transformer tends to remain constant with changes in load. Regulation is usually expressed as a percentage and is equal to

$$\% \text{ of regulation} = \frac{\text{no-load voltage} \times \text{load voltage}}{\text{load voltage}} \times 100$$

Some transformers show good regulation (a low percentage). Other transformers provide very poor regulation (a high percentage). For example, assume that the no-load voltage at a transformer output (secondary winding) is 126 V and that the load voltage (at a full-rated load) is 120 V. The percentage of regulation is $(126 - 120)/120 = 0.05; 0.05 \times 100 = 5\%$. If the load voltage is dropped to 110 V (poorer regulation), the percentage of regulation is $(126 - 110)/110 = 0.145; 0.145 \times 100 = 14.5\%$.

Transformer regulation is also expressed, and measured, as a percentage of impedance. This measure (sometimes known as "percent Z" or "$\%Z$" on transformer nameplates) involves the primary (or source voltage necessary to form rated current to flow through a short-circuited secondary (or load) winding, compared to the rated primary voltage, or

$$\% \text{ of impedance} = \frac{\text{primary voltage (with secondary shorted)}}{\text{rated primary voltage}} \times 100$$

For example, assume that a transformer with a primary rated at 240 V requires 12 V to produce the rated current in a short-circuited secondary winding. The percentage of impedance is $(12/240) \times 100 = 5\%$. If the voltage must be increased to 24 V, the percentage of impedance is $(24/240) \times 100 = 10\%$.

6-1.6 Transformer Testing

Although it is not necessarily the job of the electrical system designer to test transformers, the need for such tests often arises in practical work. The following paragraphs describe simple procedures for the test most often needed.

Checking Phase Relationships. When two supposedly identical transformers must be operated in parallel (at the primary or source) and the transformers are not marked as to phase or polarity, the phase relationship of the transformers can be checked using a voltmeter and a power source. The test circuit is shown in Fig. 6–2.

The transformers are connected in proper phase relation for parallel operation of the meter reading is zero or very low. The transformers are connected for series operation (secondaries in series, adding) if the output voltage is approximately double that of the normal secondary output of one transformer.

If the meter reading is zero or very low, but series operation is desired,

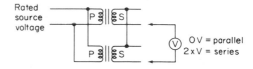

FIGURE 6-2 Checking transformer phase relationships.

reverse either the primary or the secondary terminals (*but not both*) of one transformer (*but not both transformers*).

If the meter reading is double the rated secondary voltage, but parallel operation is desired, reverse either the primary or secondary terminals (not both) of one transformer (not both transformers).

If the meter indicates some secondary voltage, far below that of the rated voltage (but not zero), it is possible that the transformers are *not identical*. One transformer has a greater output than the other. This condition results in considerable *local current* flowing in the secondary winding, and produces a power loss (if not actual damage to the transformers).

Note that the test circuit of Fig. 6-2 not only checks phase relationships but also indicates *matching* of transformers.

Checking Polarity Markings. Most transformers' windings are marked as to polarity (or phase), as discussed in Sec. 6-1.2. However, the markings may not be clear, or some new system of identification may be used. When this occurs, it is possible to identify the polarity markings using only a voltage source and a voltmeter.

When a transformer has only one primary and one secondary (or one high and one low voltage) winding, the problem of identifying the polarity markings is relatively simple. From a practical standpoint, there are only two problems of concern with single-winding transformers: (1) the relationship of the primary to the secondary, and (2) the relationship of the markings on one transformer to those on another.

The phase relationship of primary to secondary can be found using the test circuit of Fig. 6-3. First connect one primary terminal to one secondary

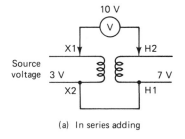

(a) In series adding

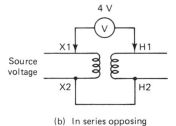

(b) In series opposing

FIGURE 6-3 Checking transformer polarity markings.

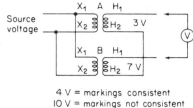

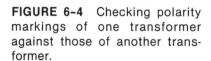

FIGURE 6-4 Checking polarity markings of one transformer against those of another transformer.

4 V = markings consistent
10 V = markings not consistent

terminal. Apply a voltage at the primary (or low-voltage winding). The voltage need not be at the full rating of the transformer. A lower voltage is often more convenient. If you have difficulty in identifying the high-voltage and low-voltage windings, remember that the low-voltage winding has larger wires to accommodate a higher current.

With a test voltage applied to the low-voltage winding, measure the voltage across X1 and H2, and then across X1 and H1. Assume that there is 3 V across the low-voltage winding and 7 V across the high-voltage winding.

If the windings are as shown in Fig. 6-3a, the 3 V is added to the 7 V, and appears as 10 V across X1 and H2. If the windings are as shown in Fig. 6-3b, the voltages oppose, and appear as 4 V (7 − 3) across X1 and H1. In either case, the phase relationship between primary and secondary is established.

The phase relationship of markings on one transformer to those on another can be found using the test circuit of Fig. 6-4. Assume that there is a 3-V output from the high-voltage winding of transformer A and a 7-V output from transformer B. If the markings are consistent on both transformers, the two voltages oppose and 4 V is indicated. If the markings are not consistent, the two voltages add, resulting in a 10-V reading.

When a transformer has multiple windings, the problem of identifying the polarity markings is more complex. Not only is it necessary to establish the phase relationship between primary and secondary windings, but it is also necessary to find phase relationships between the primary windings (as well as between secondary windings).

The phase relationship of multiwinding transformers can be found using the test circuit of Fig. 6-5, which is for a transformer with two primary (240

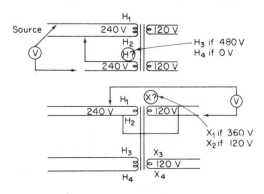

FIGURE 6-5 Checking polarity markings on multiwinding transformers.

V) and two secondary (120 V) windings. The same procedure can be used for three-phase transformers, except that one additional set of windings (primary and secondary) must be checked. The following steps describe the test procedure.

1. Connect any winding to the proper voltage according to the nameplate information, as shown in Fig. 6–5.

2. Arbitrarily assign H1 and H2 to the source winding (or X1 and X2 if the winding is of low voltage).

3. Check the other three voltages. They should be as shown in Fig. 6–5 (one 240-V and two 120-V windings).

4. Connect one lead of the other 240-V winding to H2 of source winding.

5. If the voltmeter indicates 480 V, the voltages are additive, and the top terminal is odd-numbered or H3.

6. If the voltmeter indicates 0 V, the voltages are opposing, and the terminal is even-numbered or H4.

7. If the transformer is three-phase, repeat the procedure for the third high-voltage winding (H5 and H6).

8. Repeat the procedure with each of the 120-V windings. If the voltmeter reads the sum of 240 and 120 V, or 360 V, the common terminal of the 120-V winding is X1. If the voltmeter shows 240 − 120 V, or 120 V, the terminal is X2. Terminals X3 and X4 (as well as X5 and X6) can be found in the same way.

Measuring Copper Loss.   Copper loss can be measured using the test circuit of Fig. 6–6. A wattmeter is used in the circuit of Fig. 6–6a to measure input power. If a wattmeter is not available, the circuit of Fig. 6–6b can be used. However, the circuit of Fig. 6–6b indicates the copper loss in VA (or kVA) rather than watts.

1. With either circuit, increase the variable voltage source from zero until the ammeter in the low-voltage (short circuit) winding indicates the rated current for the winding. Typically, the variable voltage source is *very low* in relation to the rated voltage for the winding.

2. For the circuit of Fig. 6–6a, simply note the wattmeter reading to determine the copper loss (in watts).

3. For the circuit of Fig. 6–6b, multiply the variable voltage indication by the primary or input ammeter reading to find the copper loss (in VA or kVA).

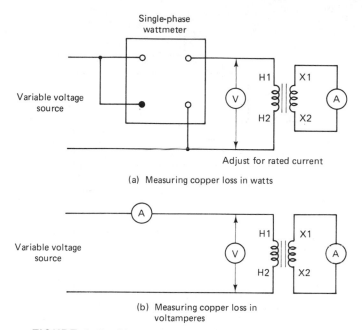

(a) Measuring copper loss in watts

(b) Measuring copper loss in voltamperes

FIGURE 6-6 Measuring transformer copper loss.

Measuring Core Loss. Core loss can be measured using the test circuit of Fig. 6–7. A wattmeter is used in the circuit of Fig. 6–7a to measure input power. The circuit of Fig. 6–7b indicates the core loss in VA (or kVA).

1. With either circuit, apply the rated voltage to the low-voltage winding. All other windings must be open circuit.

2. For the circuit of Fig. 6–7a, simply note the wattmeter reading to find the core loss (in watts).

3. For the circuit of Fig. 6–7b, multiply the voltmeter reading by the ammeter reading to find core loss (in VA or kVA).

Measuring Transformer Efficiency. Transformer efficiency is determined by adding the core loss and copper loss to the rated output to find the required input, as discussed in Sec. 6–1.4.

Measuring Percentage of Impedance. The percentage of impedance for transformers can be measured using the test circuit of Fig. 6–8.

1. Increase the variable voltage source from zero until the ammeter in the low-voltage (short-circuit) winding indicates the rated current for that winding. Typically, the variable voltage source is *very low* in relation to the rated voltage for the winding.

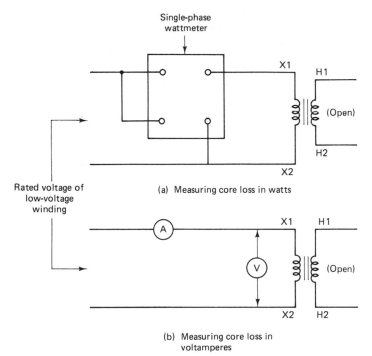

(a) Measuring core loss in watts

Rated voltage of
low-voltage
winding

(b) Measuring core loss in
voltamperes

FIGURE 6-7   Measuring transformer core loss.

2. The percentage of impedance is found when the variable source voltage is divided by the rated voltage for the high-voltage winding, and
the result is multiplied by 100. For example, assume that 6 V must
be applied to a 240-V high-voltage winding to produce an ammeter
reading (in the low-voltage winding) equal to the rated winding current. Then the percentage of impedance is $(6/240) \times 100 = 2.5\%$.

Measuring Transformer Regulation.   The transformer regulation can be
measured using the test circuit of Fig. 6–9. The required load, shown as resistance R1, must be of a value that produces full-rated current flow in the
secondary or output winding. For example, if the winding is rated at 120 V
and 6 A, the value of R1 is $120/6 = 20 \ \Omega$.

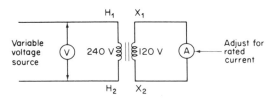

FIGURE 6-8   Measuring transformer efficiency.

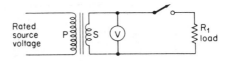

FIGURE 6-9 Measuring transformer regulation.

1. Apply the rated voltage to the input winding.

2. Measure the secondary output voltage without a load (open switch).

3. Apply the load (close the switch) and measure the secondary output voltage with the load.

4. Find the percentage of regulation using the information of Sec. 6–1.5.

Measuring Impedance Ratio. The impedance ratio of a transformer is the square of the winding ratio. Impedance ratio should not be confused with percentage of impedance.

If the winding ratio of a transformer is 15:1, the impedance ratio is 225:1 (15 × 15). Any impedance value placed across one winding is reflected onto the other winding by a value equal to the impedance ratio. For example, assume an impedance ratio of 225:1 and a 1800-Ω impedance placed on the primary. The secondary then has an impedance of 8 Ω. Similarly, if a 10-Ω impedance is placed on the secondary, the primary has a reflected impedance of 2250 Ω.

Impedance ratio is related directly to turns ratio (primary to secondary). However, turns-ratio information is not always available, so the ratio must be calculated using a test circuit as shown in Fig. 6–10.

Measure both the primary and secondary voltage. The rated voltages should be used for this test. However, lower-than-rated voltage are acceptable for test purposes.

The *turns ratio* is equal to one voltage divided by the other. The *impedance ratio* is the square of the turns ratio. For example, assume that the primary shows a 120 V, with 24 V at the secondary. This indicates a 5:1 turns ratio and a 25:1 impedance ratio.

Measuring Winding Balance. Center-tapped transformers are sometimes used in 240/120-V three-wire single-phase systems. In other cases, a transformer with two 120-V windings are used. In either case, there is always some imbalance between the windings or on both sides of the center tap. That

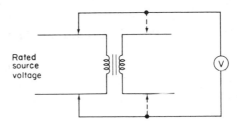

FIGURE 6-10 Measuring transformer impedance ratio.

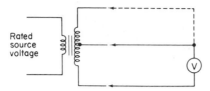

FIGURE 6-11 Measuring transformer winding balance.

is, the turns ratio and impedance are not exactly the same on both sides of the center tap (or for both windings). The imbalance is usually of no great concern unless the imbalance is severe.

It is possible to find a large imbalance by measuring the d-c resistance on either side of the center tap. However, it is usually more practical to measure the voltage on both sides, as shown in Fig. 6–11. If the voltages are equal, the transformer winding is balanced. If a large imbalance is indicated by a large voltage difference, the winding should be checked with a ohmmeter for shorted turns or a similar failure.

6-2 BASIC TRANSFORMER CONNECTIONS

When a transformer has only one primary winding and one secondary winding (or one high-voltage and one low-voltage winding), it is relatively simple to make the transformer connections. About the only concern is that the primary or input is connected to the rated voltage, and that the secondary or output is connected to a load that draws no more than the rated current. It may be necessary to convert between watts and VA (or kVA), and possibly include the effects of power factor, but these are simple calculations, as described in Chapter 9.

When a transformer has multiple windings, the connections are more difficult. The same is true when two or more transformers must be used to match a given input and output voltage. In theory, a transformer can have an infinite number of windings, with a corresponding number of connections. In practical work, a single-phase transformer generally has no more than two high-voltage windings and two low-voltage windings. Three-phase transformers usually have no more than three high-voltage and three low-voltage windings. The following paragraphs describe these basic transformer connections.

6-2.1 Single-Phase Transformer Connections

Figures 6–12 through 6–15 show typical single-phase transformer connections. In all four illustrations, the transformer is rated at 1.2 kVA, the two high-voltage windings are rated at 1200 V, and the two low-voltage windings are rated at 120 V.

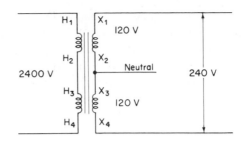

FIGURE 6-12 Single-phase transformer connections for 120/240-V three-wire.

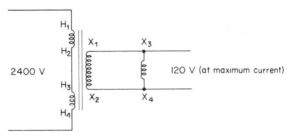

FIGURE 6-13 Single-phase transformer connections for 120-V two-wire at maximum current.

In Fig. 6–12, the source is 2400 V, and the desired output is 120/240-V three-wire. Note that the tap for the neutral wire is taken at the junction of X2 and X3. Since the transformer is rated at 1.2 kVA, the low-voltage windings carry 5 A (1200 W/240 V = 5 A), whereas the high-voltage windings carry 0.5 A (1200 W/2400 V = 0.5 A).

In Fig. 6–13, the source is also 2400 V, but the desired output is 120-V two-wire. Note that both low-voltage windings are connected in parallel. It is possible to obtain 120 V from only one low-voltage winding. However, each low-voltage winding is capable of carrying only 5 A. If the load draws the full 10 A (to provide the full 1.2 kVA), both low-voltage windings must be used.

In Fig. 6–14, the source is reduced to 1200 V, but the desired output

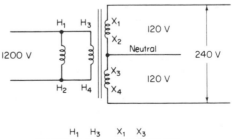

FIGURE 6-14 Single-phase transformer connections for 120/240-V three-wire (with reduced high voltage).

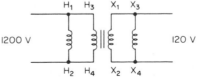

FIGURE 6-15 Single-phase transformer connections for 120-V three-wire maximum current, and reduced high voltage.

remains at 120/240-V three-wire. The tap for the neutral wire is still taken from the junction of X2 and X3. However, the high-voltage windings are connected in parallel. With a 240-V output and 1.2 kVA rating, the low-voltage output windings carry 5 A. Since there is a 5:1 voltage ratio (1200/240), the current ratio is also 5:1, and the high-voltage windings must carry a total of 1 A. This 1 A is divided between the two high-voltage windings in parallel (0.5 A in each winding).

In Fig. 6–15, the source is at 1200 V and the output is 120-V two-wire. This requires that both the high- and low-voltage windings be connected in parallel to handle the full 1.2-kVA rating.

Note: When connecting any transformer windings in parallel, always connect the odd-numbered terminals together and the even-numbered terminals together.

6-2.2 Three-Phase Transformer Connections

Figures 6–16 through 6–18 show typical three-phase transformer connections. The circuits apply to a three-phase transformer with multiple windings, or to three separate single-phase transformers. Three transformers are often used in three-phase electrical systems. One advantage is reduced transformer size. Another advantage is that a failure in one section of a multiwinding transformer usually requires replacement of the entire transformer.

The three connections shown in Figs. 6–16 through 6–18 are delta-to-wye, delta-to-delta, and wye-to-delta, respectively. It is also possible to con-

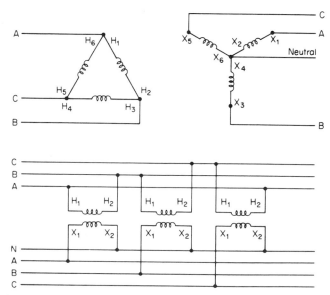

FIGURE 6-16 Three-phase transformer connections for delta-to-wye.

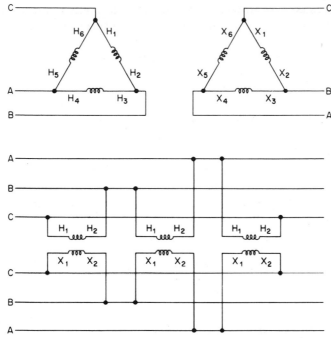

FIGURE 6-17 Three-phase transformer connections for delta-to-delta.

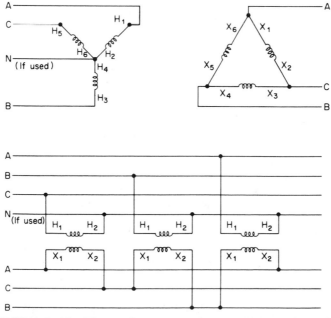

FIGURE 6-18 Three-phase transformer connections for wye-to-delta.

nect transformers wye-to-wye. However, wye-to-wye is not often used in interior electrical wiring. One problem is that a neutral wire must be used on both input and output windings. The wye-to-wye connection is found most often on high-voltage transmission lines, which are the responsibility of the utility company.

In making any three-phase connections, it is absolutely necessary to assure that the correct phase or polarity is observed for each section (primary and secondary) and *from each section to all other sections*. The procedures of Sec. 6–1.6 can be used if there is doubt as to identification on the transformer terminals.

Note that both the geometric and line versions of the three-phase connections are shown on Figs. 6–16 through 6–18. The line versions show the terminals for three-separate transformers. The geometric versions show the terminals for a single three-phase transformer with multiple windings.

6-2.3 Calculating Transformer Power Ratings and Line Currents for Three-Phase Connections

Figure 6–19 shows a typical three-phase power transformer connection. The system shown is delta-to-wye, using three identical transformers, each with a 480-V high-voltage winding and two 120-V low-voltage windings.

Assume that the source is 480-V three-phase and that the load is balanced three-phase, 60 kW, with a power factor of 0.8, and requires 120 V. Find the minimum power rating (in both watts and VA) for each transformer, the cur-

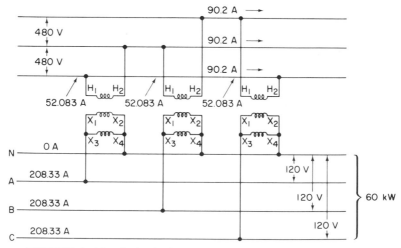

FIGURE 6-19 Calculating transformer power ratings and line currents for three-phase connections.

rent in the primary and secondary of each transformer, the input line current from the 480-V delta source, and the output line current at the wye load.

1. With a 60-kW load and three transformers, the *minimum* transformer rating in watts is 60 kW/3 = 20 kW.

2. With a 60-kW load, three transformers, and a power factor of 0.8, the minimum transformer rating in VA is 60 kW/0.8 power factor = 75 kVA; 75 kVA/3 = 25 kVA (per transformer).

3. With 25 kVA per transformer and a primary voltage of 480 V, the primary current at each transformer is 25 kVA/480 V = 52.083 A.

4. With 25 kVA per transformer and a secondary voltage of 120 V, the secondary current at each transformer is 25 kVA/120 V = 208.33 A.

5. With a total power of 60 kW, and a power factor of 0.8, the line current in the 480-V line is 60 kW/(1.732 × 480 × 0.8) = 90.2 A. Or, with a total of 75 kVA, the line current in the 480-V line is 75 kVA/(1.732 × 480) = 90.2 A.

6. With a total power of 60 kW and a power factor of 0.8, the line current in the 120/208-V (wye-connected) line is 60 kW/(1.732 × 208 × 0.8) = 208.33 A. Note that this agrees with the secondary current previously calculated.

6-3 SPECIAL TRANSFORMER CONNECTIONS

In addition to conventional single-phase and three-phase transformer connections, there are many other ways in which transformers can be connected. However, there are only three systems of particular interest: (1) autotransformer connections, (2) the Scott or T connection, and (3) the open delta connection. The following sections discuss each of these systems.

6-3.1 *Autotransformer Connections*

Practically any transformer can be connected as an autotransformer. That is, the primary and secondary windings can be connected directly to produce the desired manipulation of input and output voltages. In a conventional transformer, the primary and secondary windings are electrically isolated. When the autotransformer connection is made, some of the power is conducted directly from the source to the load. In effect, the autotransformer transforms only part of the load. This makes it possible to use a transformer of a lower power rating than normally required.

The maximum voltages available from an autotransformer are equal to the sum of the winding voltages. For example, assume that an autotransformer has one 480-V winding (high voltage), and two 120-V low-voltage windings. The outputs available are 120 V, 240 V, 480 V, 480 V + 120 V (600 V), and 480 V + 120 V + 120 V (720 V).

The autotransformer can be operated by a source voltage equal to the voltage rating for any one winding, or a combination of the winding voltages. For example, using the same autotransformer, the system can be operated from a source of 120 V, 240 V, 480 V, 600 V, or 720 V.

These voltage combinations are shown in Fig. 6-20. To use the information of Fig. 6-20, simply select the input and output terminals that match the available and desired voltages. Using the example shown in Fig. 6-20 (input 480 V, output 720 V), the input is connected at A and B; the output is taken from E and H. As shown, the input leads are connected to terminals H1 and H2, the output leads are connected to H1 and X4, and terminals H2 to X1, X2 to X3 are interconnected.

No matter what winding combination is used, the power rating of the transformer is reduced from the full rating by an amount equal to the ratio of input and output voltages. For example, if the input voltage is 480 V, and the output is 720 V (as shown in Fig. 6-20), the ratio is 2:3, and the rating is reduced by two-thirds. Assuming a load of 60 kVA, the transformer rating can be reduced to 20 kVA (60 kVA × 2/3 = 40 kVA; 60 kVA − 40 KVA = 20 kVA).

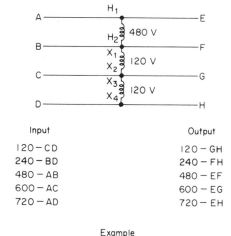

Input	Output
120 – CD	120 – GH
240 – BD	240 – FH
480 – AB	480 – EF
600 – AC	600 – EG
720 – AD	720 – EH

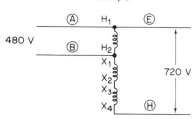

FIGURE 6-20 Basic autotransformer connections.

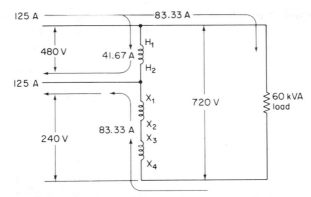

FIGURE 6–21 Example of how an autotransformer can be operated at reduced ratings.

This reduction of power rating can be proved using the calculations of Fig. 6–21. The output (or load) current is 60 kVA/720 V = 83.33 A. The input (or source) current is 60 kVA/480 V = 125 A. The current through the high-voltage winding (H1-H2) is 125 A − 83.33 A = 41.67 A. The kVA rating for the high-voltage winding is 41.67 A × 480 V = approximately 20 kVA. The current through the low-voltage windings (X1-X2, X3-X4) is 83.33 A. The kVA ratings for the low-voltage windings is 83.33 A × 240 V = approximately 20 kVA.

120-V Single-Phase from 240-V Three-Phase with an Autotransformer. One of the most common uses for an autotransformer is to tap 120-V single-phase power for lighting circuits from 240-V three-phase (used for motors or similar heavy loads). Figure 6–22 shows such a system. The 240-V supply is connected as three-phase delta. The transformer is a center-taped 240-V winding, typical of those used in 120/240-V three-phase systems. The primary winding is unused. The transformer can be a 120-V to 120-V isolation transformer.

No matter what type of transformer is used, the transformer power rat-

FIGURE 6–22 Example of autotransformer used to tap 120-V single-phase power from 240-V three-phase.

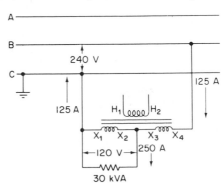

ing can be reduced to one-half that of the load. For example, assume a 30-kVA load on the 120-V output. The output current is 30 kVA/120 V = 250 A. The input current is 30 kVA/240 V = 125 A. So the current through either winding is 125 A. The kVA rating for either 120-V winding is 125 A × 120 V = 15 kVA (or one-half of the 30-kVA load).

Because of the difference in input and output currents, the transformer should be located as near the 120-V single-phase load as possible. This keeps the 250-A lines at a minimum length. Of course, the three-phase phase lines to the transformer are longer, but these lines carry only one-half the current (125 A).

Also note that one line of the three-phase supply must be grounded. *The NEC forbids use of autotransformers for lighting and appliance branch circuits unless there is a grounded conductor common to both the primary and secondary circuits.*

600-V Three-Phase from 480-V Three-Phase with an Autotransformer.   Another common use for autotransformers is to provide 600-V three-phase from 480-V three-phase. Electric motors typically operate from 240-, 480-, and 600-V three-phase sources. Many motors have dual windings and can operate from either 240 V or 480 V. However, 600-V heavy-duty motors usually have only one set of windings and require the full 600 V. It is possible to use the autotransformer principle to convert 480 V to 600 V. Such a system is shown in Fig. 6–23. Three identical transformers are required. The high-voltage windings are 480 V; the low-voltage windings are 92.5 V. The output voltage is the vector sum of the two winding voltages (600 V) between each line. The system is connected as delta at both input and output.

As a guideline, the combined power ratings for all three transformers should be approximately 25% of the load. For example, with an assumed load of 60 kVA, the power rating of each transformer should be 5 kVA (60 kVA × 0.25 = 15 kVA; 15 kVA/3 = 5 kVA per transformer).

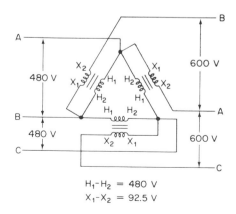

H_1-H_2 = 480 V
X_1-X_2 = 92.5 V

FIGURE 6–23 Example of autotransformer used to provide 600-V three-phase from 480-V three-phase.

6-3.2 Scott or T (Two-Phase) Transformer Connection

Practically all power distribution in the United States is either single-phase (typically three-wire 240 V) or three-phase. All electrical systems are now designed for single-phase or three-phase. However, two-phase systems may exist in certain isolated areas. In those cases, where it is not practical (or economical) to change from a two-phase system, it is necessary to convert the three-phase distribution to two-phase. Figure 6–24 shows such a system.

In Fig. 6–24, the input is conventional 480-V three-phase and the output is 240-V two-phase. This output should not be confused with conventional 240-V three-wire, or with the "open delta" system described in Sec. 6–3.3. The output of the Fig. 6–24 circuit is a pair of 240-V voltages, displaced in phase by 90°.

Generally, two transformers are used (although a special Scott transformer can be constructed). Either way, the primary windings must be tapped. One winding is center-tapped. The other winding is tapped at 0.86 of the winding voltage. With a 480-V winding, the tap is at 415 V. The transformer with the center-tapped winding is called the *main transformer*. The transformer with the 0.86 winding tap is called the *teaser transformer*.

The Scott transformer system can also be used to convert a two-phase voltage to a three-phase. This is done by applying the two-phase input at the secondaries and taking the three-phase output from the primaries. However, such an arrangement is of little practical value. (The Scott system was developed to save otherwise obsolete two-phase motors, and so on, when utility companies changed to single-phase and three-phase distribution.)

The Scott system has a power limitation in that the transformers must

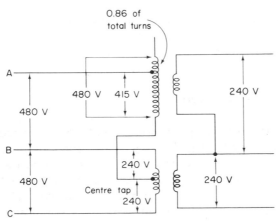

FIGURE 6–24 Example of Scott or T transformer connection to provide 240-V two-phase from 480-V three-phase.

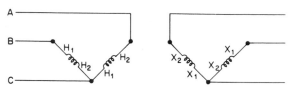

FIGURE 6-25 Open delta transformer connections.

have an approximate 116% higher rating than the load. For example, if the load is 10 kVA, the transformer must have 11.6-kVA power ratings.

6-3.3 Open Delta Transformer Connection

The open delta transformer connection is generally used only as an emergency or temporary system. As shown in Fig. 6–25, the open delta is identical to the delta-to-delta system (Fig. 6–17) except that one transformer (or one set of primary and secondary windings) is eliminated. If one transformer (or one set of windings) in a conventional delta-to-delta system becomes defective, simply disconnect the windings (both primary and secondary) of the defective transformer and leave the system with two transformers (or two sets of windings), as shown in Fig. 6–25.

With the open delta, the combined power rating of the two transformers is reduced to 57.7% of the combined power ratings of three transformers. For example, if the three transformers in a conventional delta-to-delta system are capable of handling a 10-kVA load, the load capability drops to 5.77 kVA. For this reason, open delta is rarely (if ever) used for original design.

6-4 INSTRUMENT TRANSFORMERS

Instrument transformers make it possible to measure high voltage and current with low-scale voltmeters and ammeters. There are two basic types of instrument transformers: potential or voltage transformers, and current transformers.

6-4.1 Current Transformers

Typical current transformers are shown in Fig. 6–26. One current transformer consists of a bar of conducting metal (typically copper or aluminum to match the system conductors) with several turns of wire wrapped around the conducting bar. In use, the conductor bar is connected in the line to be measured and a low-scale ammeter connected to the turns of wire. The ratio of turns between the two circuits depends on how large a current is anticipated and the capability of the meter. For example, with a maximum anticipated line current of 600 A and an ammeter capable of reading from 0 to 5 A (maximum), the

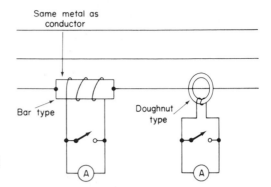

FIGURE 6-26 Typical current transformers.

turns ratio is $600/5 = 120{:}1$. Now assume that 480 A passes through the line. The ammeter indicates 4 A. In some cases, the ammeter scale is marked with two sets of numbers (one set with a current transformer and one set without).

Another type of current transformer is the "doughnut" type. This type consists of a magnetic core in the shape of a ring. The line to be measured is passed through the ring. The ammeter is connected to windings around the ring. Operation of the doughnut type is the same as for the bar-conductor type. Current transformers are also used with portable clamp-on type of ammeter.

Note that the current transformers shown in Fig. 6–26 are provided with a shorting switch. This switch must be closed (or a shorting strap placed across the current transformer terminals) when the ammeter is not connected. One reason for this short is that a high voltage is developed across the transformer secondary winding. For example, if the line is carrying 120 V and the turns ratio is 120:1, the winding voltage is $120 \times 120 = 14{,}400$ V. Of course, the ammeter acts as a short circuit and drops the voltage (when connected). With the ammeter disconnected, the high-voltage potential exists. An open secondary circuit can also result in a voltage drop in the primary circuit (which is carrying its own load).

From a design standpoint, the main concern is that the current rating of the transformer must not be exceeded. That is, always select a current transformer with a current rating at least equal to (and preferably higher than) the maximum anticipated load. Also, make certain that the current (or turns) ratio matches that of the meter (or select meters to match available current transformers).

6-4.2 Potential or Voltage Transformers

A typical voltage transformer is shown in Fig. 6–27. As shown, the voltage transformer has a high-voltage primary (connected across the high-voltage lines to be measured) and a low-voltage secondary (connected to a low-scale voltmeter).

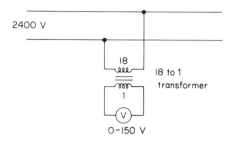

FIGURE 6-27   Typical voltage transformer.

When voltmeter is 133.33 V, line = 2400 V

The ratio of turns depends on how larger the anticipated line voltage is and the capability of the meter. For example, assume that a 2400-V line is to be measured and the meter is capable of reading from 0 to 150 V. Further, assume that the line voltage might reach 2700 V maximum. A satisfactory turns ratio is 2700/150 = 18:1. Assuming that the line remains at 2400 V, the voltmeter reads 133.33 V (unless the scale has been altered to show the actual line voltage).

As in the case of current transformers, voltage transformers are often available (in matched sets) with meters. As an alternative, the ratio information is specified on the transformer nameplate.

6-4.3   *Instrument Transformers for Wattmeters*

As discussed in Chapter 9, a wattmeter requires both voltage and current to indicate the power consumed in a system. Conventional wattmeters typically operate on voltage in the range of 120 V and at currents below 10 A. When measuring power in a high-voltage, high-current system, it is necessary to use both current transformers and voltage transformers. Such an arrangement is shown in Fig. 6–28.

Generally, each wattmeter requires one current transformer and one voltage transformer. In the circuit of Fig. 6–28, two wattmeters are required to measure three-phase power (see Chapter 9).

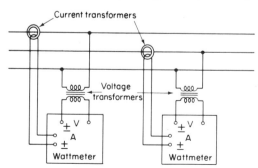

FIGURE 6-28   Instrument (voltage and current) transformers for wattmeters.

6-5 USING TRANSFORMERS IN COMMERCIAL AND INDUSTRIAL LIGHTING

The most common use for transformers in commercial and industrial wiring is in the lighting system. Of course, there are exceptions to this rule. For example, transformers provide the only practical means for electrical power where a particular load requires a particular voltage and all other loads in the system require a different voltage.

Most lighting systems require 120 or 277 V (also referred to as 115 and 265 V). If these voltages are not readily available at the service entrance or it is not economical to receive these voltages from the utility company, transformers are used to provide the voltages. For example, assume that an industrial building requires 480 V for electrical motors. The same building requires either 120 or 277 V (or both) for lighting. In some cases it is more economical for the utility company to supply only one voltage. In other cases the long conductors from the service entrance to the lighting, combined with the relatively low voltage (120 or 277 V compared to 480 V), result in very high currents and large conductors. If power can be distributed from the service entrance to the lighting in three-phase form, and then converted to single-phase, the current (and conductor size) can be reduced (because the power in three-phase is increased by 1.732 compared to single-phase).

6-5.1 Typical Transformer Connections for Commercial and Industrial Lighting

Figures 6–29 through 6–33 show the most common arrangements for lighting transformers. Figure 6–34 shows a typical commercial/industrial transformer assembly.

In Fig. 6–29, a three-phase delta is transformed into three separate sources of 120/240-V three-wire power for distribution to lighting.

In Fig. 6–30, two transformer primaries are connected to the three-phase

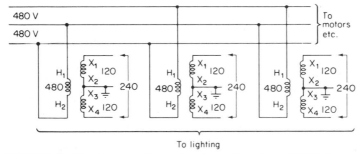

FIGURE 6-29 Connections for transforming 480-V three-phase delta into three separate 120/240-V three-wire sources.

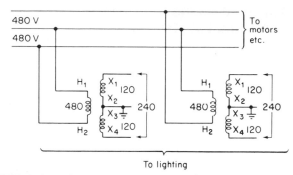

FIGURE 6-30   Connections for transforming 480-V three-phase delta into two separate 120/240-V three-wire sources.

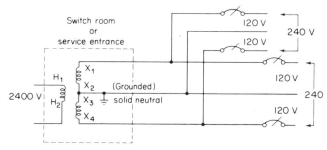

FIGURE 6-31   Connections for transforming 2400-V into 120/240-V three-wire power, distributed into two separate loads.

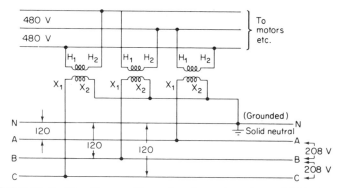

FIGURE 6-32   Connections for transforming 480-V three-phase delta into 120/240-V three-phase four-wire sources.

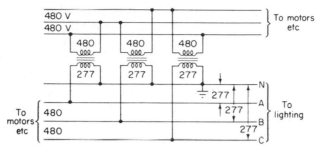

FIGURE 6-33 Connections for transforming 480-V three-phase delta into 277/480-V three-phase four-wire wye distribution.

FIGURE 6-34 Typical commercial/industrial transformer assembly. (Courtesy Square D Company.)

supply in an arrangement similar to the open delta (Sec. 6–3.3). This results in two sources of 120/240-V three-wire power for lighting.

In Fig. 6–31, a high-voltage source is converted to a single source of 120/240-V power. However, the power is distributed to two separate loads.

In Fig. 6–32, a three-phase 480-V source is transformed into a 120-/208-V three-phase four-wire system. In the circuit of Fig. 6–32, the 120 V (line to neutral) is available for lighting, with 208 V (line to line) available for other loads. A neutral conductor is also available for unbalanced loads.

Figure 6–33 shows a three-phase delta system transformed into a three-phase wye four-wire at 277/480 V. This system is the most common for lighting in large industrial buildings, large offices, and the like.

6-5.2   Design Notes for Lighting Transformer Systems

The following notes investigate the capabilities and limitations of various lighting transformer systems for commercial and industrial use.

Grounding.   No matter what system is used, *the NEC requires a ground or neutral conductor for all lighting.* A ground is required for the safety reasons discussed in Chapter 5. A neutral is required where there is a possibility of unbalanced loads. Any electrical system should be designed so that the loads are balanced. Each phase of three-phase (or both sides of 120/240-V three-wire) should have an equal load (as near as practical). So if there are 100 lamps in a lighting system, 33 or 34 lamps should be connected to each phase of three-phase distribution (or 50 lamps should be connected to each side of a 120/240-V three-wire system). Under these conditions, the ground or neutral current is zero (or nearly zero). However, there is no guarantee that all lamps are used simultaneously (except in special cases). So it is possible for loads to be unbalanced even in well-designed electrical wiring systems.

Placement of Transformers.   There are two schools of thought on transformer placement. One system of power distribution involves the installation of high-voltage three-phase or single-phase feeders throughout the building. The 120-V lighting loads (and any other loads requiring 120 V) are supplied by *transformers located near the load,* as shown in the distribution system of Fig. 6–29. With this system the load is conducted over longer distances by smaller conductors carrying smaller currents. Some manufacturers of electrical equipment provide a transformer and panelboard assembly complete with connecting ducts that can be secured to columns of industrial buildings. When using the system of Fig. 6–29, *conductors of the same size must be used to the last transformer,* and all conductors must be protected by overcurrent devices of the same ampacity.

The distribution system of Fig. 6–30 can also be arranged with the transformers near the loads. The main disadvantage of the open delta system of Fig. 6–30 is the reduction of power. The advantage is that two transformers are required instead of three.

An alternative to the systems of Fig. 6–29 and 6–30 is a system in which all feeders originate in a switch room or at the service entrance, as shown in Fig. 6–31. The utility company provides separate service for both the power and lighting, or the customer maintains its own transformers in the area. In a large building (either commercial or industrial) this alternative system usually results in long feeders and excessive voltage drops in the 120/240-V system. The alternative system is acceptable in smaller buildings.

Separate Services. If the load is less than about 50 kW, it is generally practical for the utility company to supply two (or more) separate services (three-phase for high power and 120/240-V three-wire for lighting and small loads). When the total power requirements are greater than 50 kW, the utility company can usually furnish power at a lower rate per kilowatthour only if one type of service is provided through one meter. However, the subject of service entrances must be checked with the local utility company.

Three-Phase 120-V Lighting. The distribution system of Fig. 6–32 can be used in place of the usual three-wire 120/240-V system in some buildings (in some larger office buildings). The service equipment is smaller with the three-phase four-wire system shown in Fig. 6–32. If this is the only service connection to the building, the utility company usually provides the transformers (outside the building).

With the system of Fig. 6–32, it is not necessary to distribute all three phases to each office. Instead, three lines (neutral and two conductors) are distributed to each office, usually through a separate meter. The lighting and small appliances are operated from the 120 V. Large appliances must operate from 208 V. This is the main disadvantage of the 120/208-V system (all appliances are not always available in 208 V). In the case of some office buildings, many occupants may have 240-V appliances (particularly large room air conditioners), which creates an obvious problem.

Three-Phase 277/480-V Lighting. The distribution system of Fig. 6–33 can be used most effectively where large electric motor loads are combined with heavy lighting requirements. This is typical of large industrial plants. Again, if this is the only service connection to the building, the utility company usually provides the transformers, and the input to the service entrance appears at three-phase four-wire 277/480 V.

The use of 277-V lighting units permits a branch circuit 2.5 times longer than a 120-V branch circuit. This is an important advantage in large buildings, particularly those with high ceilings. Another advantage of the 277/480-V system is the availability of three-phase for motors at any point throughout the system. *The NEC permits tapping conductors of at least one-third the ampacity of the larger, if the smaller conductors do not extend more than 25 ft.* For example, assume that the main conductors running throughout the system are 100 A. Then 33-A conductors can be used to tap power for electric motors (or other loads), provided that the 33-A conductors do not extend over 25 ft. The smaller conductors must be suitably protected against mechanical injury (Chapter 2) and must terminate in 33-A (or less) circuit breakers that control the electric motors.

The main disadvantage of 277/480-V distribution is that only large-base lamps can be used for lighting. This limits the choice of lamps. Lamp bases of any size can be used with 120 V. Also, if 120-V lighting must be used somewhere in the building, a separate transformer (or transformers) must be provided.

The advantages of 277-V lighting over 120-V lighting are lower currents, smaller conductors, and a higher permitted voltage drop (with the same percentage of voltage drop). In practical terms this means that 277 V is better for long runs than 120 V. As a guideline, if the branch circuits must be longer than 100 ft, use 277 V for all lighting (if possible). If the branch circuit run is between 50 and 100 ft, weigh all factors (cost, convenience, etc.) carefully. Generally, if *all lighting* can be 277 V, and the run is between 50 and 100 ft, use 277 V. If the branch circuit run is less than 50 ft, there is probably little advantage in the 277-V distribution. An exception to this is where there is a very heavy lighting load (such as a large arena with thousands of lights) and all (or most) of the lighting can be 277 V.

7

ELECTRICAL LIGHTING DESIGN AND CONTROL

In this chapter we discuss design of electrical lighting from the standpoint of providing a given amount of light for a given area, and of selecting the proper control devices for an electrical lighting system. Commercial and industrial lighting systems are most efficiently controlled by *contactors*. Many of the contactors described in Chapter 8 for electric motor control can also be used to control lighting systems. Similarly, some (but not all) lighting contactors can also control motor loads, as well as straight resistance loads.

We discuss rule-of-thumb calculations to determine how many lamps are required to provide a given illumination level in a given interior space, and how the lamps should be arranged to provide uniform illumination throughout the space. With the number of lamps (of given wattage or voltage and current rating) established, the electrical wiring system (conductor size, voltage drop, transformer connections, etc.) for the lamps can be calculated using the information of Chapters 1 through 6. At that point in design, contactors of the proper type and rating can be selected using the information of Sec. 7-4. The selection process for both the light sources and contactors, as described here, involves the use of tables, charts, guidelines, and so on, rather than complex calculations.

There is considerable disagreement among lighting experts as to how much light is required for a given situation. In many cases the amount of light is a matter of preference. In any event, there are many charts and tables available from lighting equipment manufacturers, and various engineering societies, which show their recommendations for the amount of light in given areas and situations. This information is not duplicated here. Instead, we concentrate on how to provide a given amount of light, in a given area, under a given set of conditions, and how to select the appropriate contactor to handle the lighting load.

7-1   ELECTRICAL LIGHTING BASICS AND DEFINITIONS

Before going into practical design, let us review the basics of electrical lighting, such as the definitions and terms used for most common light sources, efficiency factors, and depreciation factors.

7-1.1   Lamps and Luminaires (Lighting Fixtures)

Lamps are held by *luminaires*. Luminaires may be functional or decorative (or both) and are usually referred to by the general public as *lighting fixtures*. In addition to holding the lamps, luminaires function to distribute and diffuse the light from the lamps. All luminaires have certain efficiency and depreciation factors. That is, in an effort to distribute and diffuse the light so that glare is minimized and lighting is uniform, luminaires cannot be designed for 100% efficiency. (A single unshaded lamp hanging from a cord in the center of a room is nearly 100% efficient, but the glare makes such an arrangement impractical.)

7-1.2   Efficiency of Luminaires

Many factors affect the efficiency of luminaires. A few of the important factors are construction of the luminaire, the type of light source, reflection from surrounding areas, cleanliness of the light sources over a period of time, and the distances from the light source to the *work plane* (the point where the level of light intensity is to be measured). Distance is a very important factor. Light varies inversely as the square of the distance from the light source. For example, if the distance doubles, the light is reduced to one-fourth.

7-1.3   Coefficient of Utilization

Many of these foregoing factors, and others, are combined into a single efficiency factor called the *coefficient of utilization* (CU). The CU for a particular luminaire is determined by experiment and is given by the luminaire manufacturer in its catalog. Typically, the manufacturers' publications include charts that give consideration to the relative light-absorbing qualities of the ceiling, walls, floors, and the luminaire. Also shown is the relative pattern of light, illustrating what percentage is reflected up to the ceiling or down to the floor.

7-1.4 Lumens and Footcandles

Lumens and *footcandles* are the most common terms used in the measurement of light. Some manufacturers, and some lighting specifications, also use the terms *footlamberts, candelas per square inch, reflectance, transmittance,* and *absorbance*. All of these terms are defined briefly in the following paragraphs.

The *candela* (pronounced can-*del*-a, and formerly known as candle) is the unit of luminous intensity of a light source in a specified direction. The standard candela (cd) is defined as 1/60 the intensity of a square centimeter of a black-body radiator operated at the freezing point of platinum (2047°K).

The *lumen* is used to measure the luminous output of lamps and luminaires. The lumen (lm) is defined as the rate at which light falls on a 1-ft^2 surface area that is equally distant (1 ft) from a source with an intensity of 1 cd. Lamps are commonly rated by their *total lumen output*. Also, the number of *lumens per watt* represents what is called the efficiency of the light source.

Illumination on a surface is measured in *footcandles*. A *footcandle* (fc) is defined as the density of light striking each and every point on a segment of the inside surface of an imaginary 1-ft radius sphere with a 1-candlepower source at the center, as shown in Fig. 7-1. Using another definition, 1 fc is the illumination on a 1-ft^2 surface over which 1 lm is distributed evenly. This means that 1 fc equals 1 lm/ft^2. The footcandle is the key unit in engineering calculations for lighting installation. (*If you want to reduce the calculations for the required amount of light for any task in any area, simply remember that 100 footcandles will probably do the job.*)

Where the metric system is used, the imaginary spherical surface is given as 1 m^2, and the radius from the light source is 1 m. When 1 lm is distributed evenly over the square meter, the illumination is said to be 1 *lux*. The ratios among candelas, lumens, footcandles, and lux are given in Fig. 7-1. The terms "lumen" and "footcandles" are used throughout this chapter.

Light is invisible as it travels through space. What the eye actually sees is brightness, the result of light being reflected or emitted by a surface directly into the eye. There are two common units of brightness: *footlamberts* and *candelas per square inch*. A surface that emits or reflects light at the rate of 1 lm/ft^2 of area has a brightness of 1 footlambert (fL), as viewed from any direction. One candela per square inch equals 452 fL. These relationships are shown in Fig. 7-2.

Reflection factor (*reflectance*), transmission factor (*transmittance*), and aborption factor (*absorptance*) are values that indicate the amount of the total light on the surface that is reflected, transmitted, and absorbed. For example, assume that 100 lm strikes a translucent spherical surface of 1 ft^2 as shown in

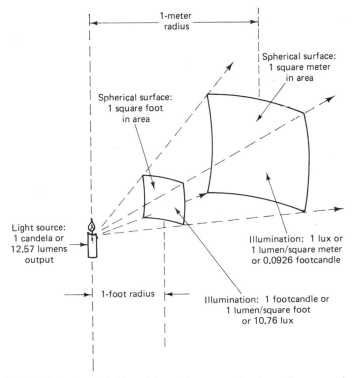

FIGURE 7-1   Relationship of lumens, footcandles, candelas, and lux.

Fig. 7-2. The incident illumination is said to be 100 fc (which is our standard light for all situations). If 60 lm of light is reflected, the brightness on the surface on the reflected light side is 60 fL, and the reflection factor is 0.6 or 60%. If 30 lm is transmitted through and emitted, the brightness on the other side of the surface is 30 fL. The transmission factor is 0.3 or 30%. Since 60 lm is reflected and 30 lm is transmitted, 10 of the total 100 lm are absorbed. So the absorption factor is 0.1 or 10%.

7-1.5   Some Common Lighting Definitions

The following terms are in common use throughout the lighting profession:

Ballast: A device used with an electric discharge lamp (fluorescent and HID) to obtain the necessary electric circuit conditions (voltage and current) to start and operate the lamp.

Brightness: In common usage, brightness is the intensity of the sensation that results from viewing a surface or space from which light comes into the eye. The footlambert is a measure of brightness.

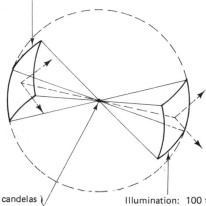

Illumination: 100 footcandles or
100 lumens/square foot;

Reflectance: 0.6 or 60%

Reflected brightness: 100 footcandles
X 0.6 reflectance = 60 footlamberts
or 60 lumens/square foot reflected

Light source: 100 candelas
or 1257 lumens output

Illumination: 100 footcandles or
100 lumens/square foot

Transmittance: 0.3 or 30%

Transmitted brightness: 100 footcandles
X 0.3 transmittance = 30 footlamberts
or 30 lumens/square foot transmitted

FIGURE 7-2   Relationship of footlamberts, reflectance, and transmittance of light.

Coefficient of utilization: A ratio of the light delivered from a luminaire to the work surface compared to the total light output emitted by the lamp. This ratio changes with room conditions and configuration of the fixture.

Contrast: The relationship between the brightness of an object and its immediate background. An example of this is the relationship between the letters printed on this page and the page itself. An example of poor contrast is a third or fourth carbon copy of a computer printout.

Diffuser: A device commonly put on the bottom and sides of a luminaire to redirect or spread the light from a source. Diffusers are used to control the brightness of the source and, in many cases, the direction of light emitted by the luminaire.

Footcandle/illumination: A measure of light striking a surface 1 ft² in area on which one unit of light (lumen) is uniformly distributed. Footcandle/illumination is a measure of the quantity of light falling on a task or work surface per unit of area.

Glare: Excessive brightness. There are two types of glare: direct and reflected. *Direct glare* occurs when a source of brightness, such as an exposed lamp, is in the line of vision. *Reflected glare* occurs when brightness from the source is reflected on a shiny surface in the line of vision.

Lamp: A term that refers to light sources that are commonly called bulbs (incandescent) or tubes (fluorescent).

Lens: A glass or plastic shield that covers the bottom, and sometimes sides, of a luminaire to control the direction and brightness of the light as it comes out of the luminaire.

Louver: A series of baffles, arranged in a geometric pattern, used to shield a lamp from view at certain angles to avoid glare from the bare lamp.

Luminaire: A complete lighting unit that consists of one or more lamps (and ballast, if needed) together with other parts designed to distribute light, position and protect the lamps, and connect the lamps to the power sources.

Nonuniform lighting: A system that has lighting located with respect to the tasks, so more lighting falls on these tasks than on surrounding areas.

Reflector: A device used to redirect the light from a lamp or luminaire by the process of reflection.

Task lighting: The lighting, or amount of light, that falls on a given viewing task or object.

Veiling reflection: Reflection of light from an object or task that partially or totally obscures the details to be seen by reducing the contrast between the object or task and its background.

7-2 LIGHT SOURCES AND LIGHTING LOADS

From a lighting standpoint, there are three types of light sources: *incandescent, fluorescent,* and *high-intensity discharge* (HID). However, from a wiring design standpoint, there are only two types of lighting loads: *tungsten load* and *ballast load.* Thus the contactors used to control electrical lighting in commercial and industrial installations are rated as to their tungsten and ballast control capacities. (In some cases, the contactors also have *resistance load* and *motor load* ratings, as discussed in Sec. 7–4.) So before we get into lighting design and contactor selection, let us discuss the loads presented by the light sources, as well as the types of light sources.

Note that this chapter describes light sources manufactured by General Electric Company. Their line is chosen since the line represents state-of-the-art equipment suitable for commercial and industrial lighting. Figure 7–3 shows some typical light sources and the related characteristics (lamp watts, initial lumens, rated life hours, and lumens per watt). Keep in mind that Fig. 7–3 shows a typical cross section of light sources. Full details can be found in the latest General Electric Company catalog.

Lamp type	Lamp (watts)	Initial (lumens)	Rated file (hours)	Lumens per watt
Incandescent	200	4,000	750	20
Fluorescent	40.7	3,250	12,000–20,000	83
Mercury vapor	400	23,000	16,000–24,000	63
Self-ballasted mercury vapor	450	14,500	16,000	
Metal halide	400	34,000	7,500–15,000	115
High-pressure sodium	400	50,000	20,000–24,000	140
Low-pressure sodium	180	33,000	18,000	183

FIGURE 7-3 Characteristics and comparative efficiencies (in lumens per watt) of selected lamps.

7-2.1 Tungsten Lighting Loads

A tungsten lighting load consists of incandescent lamps that produce light by heating a metallic (tungsten) filament to intense heat by passing an electric current through the filament. This tungsten filament is enclosed in either a vacuum envelope or one filled with an inert gas. Tungsten lighting loads include *incandescent tungsten lamps, incandescent iodine lamps, quartz-iodine lamps,* and *infrared lamps.*

The incandescent lamp can be considered the basic light source because of the common usage. Incandescent lamps also have the poorest efficiency (lowest lumens per watt rating, Fig. 7-3). The popularity of the incandescent lamp is due to the simplicity with which it can be used, and the low price of both the lamp and fixture. Also, incandescent lamps require no special equipment, such as ballasts. The most common types of incandescent lamps are: the A or arbitrary bulb-shaped lamp; the PS or pear-shaped lamp; the R or reflector lamp; the PAR or sealed-beam lamp; and the tungsten-halogen lamp.

The *tungsten-halogen lamp,* like the other incandescent lamps, uses a tungsten filament as the light source. Unlike the others, however, a family of elements known as halogens is put into the lamp. The halogens prevent lamp walls from darkening as quickly as those of other incandescent lamps, so more light is available to the task or work surface. The light output does not drop off as rapidly as the light output of other incandescent lamps.

The efficiency of incandescent lamps increases as lamp wattage increases. This makes it possible to save on both energy and fixture costs whenever you can use *one higher wattage incandescent lamp instead of two lower wattage lamps.* For example, one 100-W GS (general service) lamp produces more light (1740 lumens) than two 60-W GS lamps (860 lumens each for a total of 1720 lumens).

The specific type of incandescent lamp used, and the kind of fixture involved, also make a difference in efficiency. For example, a 75-W ellipsoidal reflector lamp delivers more light in a stack-baffled downlight than a 150-W R-40 lamp.

Incandescent Tungsten Lamps.   Tungsten lamps have a positive resistance characteristic (resistance to the flow of electric current increases as the operating temperature rises), and show an increase in resistance when the lamp is energized. Since the cold filament resistance is approximately $\frac{1}{12}$ to $\frac{1}{18}$ that of the hot resistance, the initial or *inrush* current is theoretically 12 to 18 times the normal operating current. In practice, the actual current of an incandescent tungsten lamp is slightly less than the theoretical value. Inrush is maximum during the first peak of applied voltage, and decreases to the normal operating current in a very short time. As far as the human eye is concerned, the incandescent lamp goes on instantly. Once the lamp is on and the filament heated, the resistance increases rapidly to reduce current flow. Figure 7–4 shows a comparison of typical inrush characteristics between tungsten lamp and motor (Chapter 8) loads.

Because of the high inrush problem, the contacts of control devices for lighting systems must be capable of handling high currents, even though the currents are momentary. Switches (often called *T-rated switches*) can be used in very simple tungsten load systems. However, the contactors described in Sec. 7–4 provide the most practical means for control of commercial and industrial lighting systems, both tungsten load and ballast load. When a con-

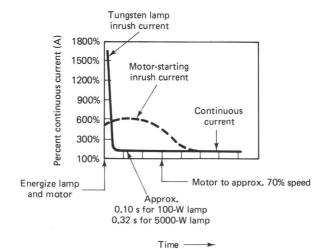

FIGURE 7–4   Comparison of typical inrush characteristics between tungsten lamps and motors.

tactor of the proper rating for a given lighting load is selected, the contacts of that contactor are capable of handling both the inrush and constant currents involved, thus eliminating the need for further design calculations.

Incandescent lamps can be operated in practically any environment. No auxiliary equipment (ballasts, transformers, etc.) is required. The load presented by incandescent is purely *resistive,* eliminating the need for power factor correction required by ballast loads. The resistive load makes power and current calculations for incandescent lamps relatively simple (current is found when wattage rating is divided by voltage).

Incandescent Iodine Lamps. During operation of tungsten filament lamps, tungsten particles evaporate from the filament and become deposited on the inside of the lamp walls. One point on the filament wears out first and the filament breaks, ending the lamp's life. If the particles of tungsten can be returned to the original place in the filament, the life of the lamp can be increased. Iodine lamps are similar to standard incandescent tungsten lamps, except that iodine gas is placed inside the lamp envelope to prolong tungsten filament lamp life and to prevent darkening of the inside lamp walls.

When the tungsten particles evaporate, they combine with the iodine gas to form tungsten iodide. The tungsten particles are redeposited on the filament whenever the tungsten iodide gas comes in contact with the filament. This cycle is repeated throughout the life of the lamp and produces an efficient self-cleaning process. Iodine lamps have approximately the same characteristics as tungsten lamps, although some have slightly lower inrush, since an iodine lamp can use a more efficient filament than a standard lamp and still obtain the same life.

Quartz-Iodine Lamps. Quartz-iodine lamps are similar to incandescent iodine lamps, except with quartz rather than glass envelopes. Quartz can withstand heat to a much greater degree than glass, and is usually used in incandescent lamps of relatively high wattage, or to decrease the physical size of the standard lamp. Quartz-iodine lamps have the same operating characteristics as incandescent iodine lamps, except that some of the high-wattage types are often preheated by a bias voltage to reduce the inrush current. In some cases, quartz is used for the envelope of tungsten-halogen lamps.

Infrared Lamps. These lamps have tungsten filaments and current characteristics very similar to general-purpose incandescent lamps, except that the filament is operated at a lower temperature in order to emit an increased number of infrared rays. The cold filament resistance is normally $\frac{1}{7}$ to $\frac{1}{10}$ that of the hot filament, resulting in theoretical inrush currents from 7 to 10 times the operating current. Inrush currents for infrared lamps last for approximately the same time period as for standard incandescent lamps.

7-2.2   Ballast Lighting Loads

A ballast lighting load consists entirely of electric-discharge (vapor) lamps such as fluorescent or mercury vapor (MV) lamps. All types of vapor lamps have a negative resistance characteristic. The resistance within the lamp decreases with an increase in current, and vice versa. Without some form of current-limiting device in the electric circuit, the current rises quickly until lamp failure occurs. This current-limiting device is known as the *ballast*. In simple terms, a ballast is an impedance used to stabilize the current in a vapor lamp, and has the property of increasing in resistance as current through the ballast increases, and vice versa. This characteristic offsets the negative resistance characteristic of the lamp, and tends to maintain a constant current through the lamp.

Several factors must be considered when designing for ballast lighting loads. The ballast and any auxiliary equipment (transformers, starters, capacitors, etc.) recommended in the lamp manufacturer's catalog *must be used*. Generally, the ballast and auxiliary equipment are part of the luminaire rather than the lamp. Thus the recommended type of luminaire *must be used*. Since a ballast is an inductance, the load is not pure resistive, and a power factor (often as low as 0.6) must be considered. (Power factor is discussed in Chapter 9.)

In some cases, the luminaires for ballast loads have some power factor correction device (such as a capacitor) which raises the power factor from a typical 0.6 to an 0.9. [These are called high-power-factor (HPF) luminaires by some manufacturers.] The power factor of the luminaire must be known when determining the current of a ballast lighting load (to determine the conductor size, voltage drop, etc.). As discussed in Chapter 9, converting wattage to current, with a power factor, is a tedious task. However, the conversion process can be reduced to a simple rule-of-thumb process, as described for fluorescent lamps next.

Fluorescent Lamps.   Fluorescent lamps (the second most common light source) are tube-shaped and contain two electrodes and mercury vapor. The inner walls are coated with fluorescent powders. When power is applied, an arc is produced by current flowing between electrodes through mercury vapor. The mercury vapor produces very little visible light, but emits ultraviolet radiation, which excites the phosphors (fluorescent powders) to produce light. Inrush current does not exceed operating current (except for some older-type preheat lamps which have an inrush of 150 to 200% of continuous operating current).

Fluorescent lamp sizes range from 4 W to about 215 W. The efficiency (lumens per watt) of a fluorescent lamp increases with lamp length (typically up to 8 ft). The reduced-wattage fluorescent lamps introduced in the last few years use from 10 to 20% less wattage than conventional fluorescent lamps, depending on size. For most commercial/industrial applications, the cool-white and warm-white fluorescent lamps provide very acceptable color and energy efficiency ratings.

Like the incandescent lamp, the fluorescent lamp can be operated to produce lower light levels. This is done with special ballasts and controls for dimming from 100% to 10% of output, or by multilevel ballasts that step down light output to specific levels (75%, 50%, etc.). Note that dimmers and other similar light-level controls for incandescent lamps cannot be used on fluorescent lamps (or other ballast lighting loads), and vice versa.

Fluorescent lamp life is rated according to the number of operating hours per start (for example, 20,000 hours at 3 hours of operation per start). The greater the number of hours operated per start, the greater the lamp life. However, because fluorescent lamp life ratings have increased, the number of times you turn a lamp on or off has become less important than it was a few years ago.

Fluorescent lamps have many design advantages over incandescent lamps. Fluorescent light is distributed over a larger lamp surface area. Fluorescent lamps produce more lumens per watt than incandescent lamps. In properly designed luminaires, fluorescent lamps can distribute their lumen output in a manner that approaches the ideal situation (where an entire ceiling area emits light evenly). Fluorescent lamps have certain disadvantages. One of the drawbacks is that fluorescent lamps (unlike incandescents) must be suitably protected (in separate enclosures) against humidity and temperature extremes.

There are two basic types of fluorescent lamps. One type has two contact pins in each end. The other type has one contact pin at each end. So there are two basic types of fluorescent lamp sockets. Both types require auxiliary equipment, such as transformers, ballasts, starters, and capacitors, all of which are usually part of the luminaire. The capacitor serves two purposes and is of particular importance in design.

One purpose of the capacitor is to minimize the *stroboscopic effect* of fluorescent lamps. The capacitor in series with one lamp enables the current in that lamp to be out of phase with the other lamp. This causes the lamps to go on and off at different times with respect to the a-c input voltage. If all fluorescent lamps in a luminaire operate together, without phase shift, a stroboscopic effect would be noted on moving objects. (This is no problem with incandescent lamps since their filaments continue to glow when alternating current passes through zero.)

The other purpose of the capacitor is to increase the power factor. As discussed, the ballasts used with fluorescent lamps are highly inductive and, without a capacitor, produce a power factor of about 0.6. With the capacitor, the power factor is increased to about 0.9.

Fluorescent lamps are not rated by voltage. Instead, the transformers and ballasts are designed for specific voltages. The most frequently used voltages for fluorescent luminaires are 120, 208, 240, and 277 V (listed in some catalogs as 115, 200, 230, and 265 V). *The NEC requires that all fluorescent luminaires be plainly marked with their voltage ratings, including ballasts and transformers. The frequency must also be included.*

Finding fluorescent currents when wattage is known. As discussed, fluorescent lighting requires more consideration when current and power are being calculated. Fluorescent lamps are generally listed in catalogs by wattage. This wattage can be used to determine the required number of lamps for a given lighting task (as discussed in Sec. 7–3). However, the wattage cannot be used directly (as is the case with incandescent lamps) to determine the wiring design (conductor size, voltage drop, etc.) because of the power factor.

If the wattage of the fluorescent lamps is known, but the current of the luminaire is not known, the approximate current (suitable for simplified wiring design purposes) can be found as follows.

Add the wattage of all lamps in the luminaire. Add 20% of the wattage to the total. Multiply the luminaire voltage by 0.9 (for a high-power-factor unit) or by 0.6 (for an uncorrected unit). Of course, if the luminaire power factor is given, use that figure. If you have no hint as to power factor, use 0.75. Divide the total wattage by this factor.

For example, assume that four 80-W lamps are used in a fluorescent luminaire, and that the voltage is 120 V. Find the current rating for both uncorrected and high-power-factor luminaires; 80 W × 4 = 320 W; 320 W × 0.20 = 64 W; 320 W + 64 W = 384 W; 384 W/120 × 0.9 = approximately 3.55 A for high-power-factor units; 384 W/120 × 0.6 = approximately 5.33 A for uncorrected units; 384 W/120 × 0.75 = approximately 4.2 A where the power factor is unknown.

Fluorescent Lamp Ballasts.   The three basic fluorescent lamp ballast types are *preheat* (requires a starter for each lamp), *rapid start,* and *instant start.* A fourth type, *trigger-start* ballasts, do not require starters, and are used with preheat lamps rated 20 W or less. We will not go into ballasts here since, from a wiring design standpoint, you must use the ballast recommended for the particular lamp. However, keep in mind that you must also use the ballast load rating of a contactor (Sec. 7–4), rather than the tungsten or resistance ratings, when selecting a contactor for control of fluorescent lighting.

Fluorescent Lamp Voltage Ratings.   Fluorescent lamps operate satisfactorily within a range of about 6% of the rated voltage. For example, if the voltage is increased to about 6% above the normal rated voltage, the wattage increases to about 108% of normal, whereas the lumens output increases to about 106%. Decreases in operating voltage cause corresponding decreases in both wattage and lumens. This factor must be considered in design of electrical distribution systems, particularly with regard to voltage drop in conductors to the luminaires. *Any system that produces a voltage drop greater than about 5% of the normal rated voltage for the lamp (or luminaire) is not satisfactory.*

High-Intensity-Discharge (HID) Lamps. HID is the term commonly used to designate four distinct types of lamps that actually have very little in common: *mercury vapor* (MV), *metal hallide* (MH), *high-pressure sodium* (HPS), and *low-pressure sodium* (LPS). Each type requires a few minutes (one to seven) to come up to full output. Also, if power to the lamp is lost or turned off, the arc tube must cool to a given temperature before the arc can be re-struck and light produced. Up to 7 minutes (for mercury vapor lamps) may be required.

In addition to the warm-up and restrike times, HID lamps have many other factors that affect lighting design (color rendition, efficiency, etc.). These factors are discussed in the following paragraphs and in Sec. 7–3. However, from a wiring design standpoint, keep in mind that HID lamps are *ballast loads,* and must be given the same consideration as fluorescent lamps (using the recommended type of ballast, considering the power factor, and using ballast load ratings of contactors).

Mercury Vapor (MV) Lamps. MV lamps consist of an envelope containing mercury vapor, and two or more electrodes. Typically, the lamp consists of two glass envelopes; an inner envelope in which the arc is struck, and an outer or protective envelope. Light is generated by the passage of an electric current through this mercury vapor atmosphere. The electrodes deliver electrical power for starting and maintaining an arc discharge through the vaporized mercury. The starting current is typically around 150% of operating current, and warm-up time is approximately 4 minutes, during which inrush drops almost linearly.

The MV lamp is available with or without a phosphor coating on the inner surface of the outer envelope. However, the lamp with the clear outer envelope has very limited application because of the blue-green light emission. A coating permits phosphors to absorb the ultraviolet component of the arc emission, and convert the component to a longer wave (to produce visible light). The selection of the phosphor mix determines the color of the light output. Manufacturers blend their phosphors to gain particular color rendition qualities for specific applications.

The MV lamp requires a ballast specifically designed for MV properties. Fluorescent lamp ballasts are not suitable for MV lamps. Also, special ballasts are required for dimming MV lamps.

MV lamps have found greatest use in industrial applications and outdoor lighting because of their low cost and long life (16,000 to 24,000 hours). Although the color-rendering qualities of the MV lamp are not as good as those of incandescent and fluorescent lamps, development of the phosphor-coated MV lamps has enabled lighting designers to use this type of HID lighting for many indoor applications, particularly in lobbies, hallways, retail display areas, and many others. MV lamp sizes range from 40 to 1000 W.

Metal Halide (MH) Lamps.   MH lamps, similar to the MV lamp in construction and operation, represent a further development of mercury vapor as a light source. The major difference is that the arc tube contains halides of various metals, in addition to mercury, that vaporize in the arc stream. Metal halide lamps are available in clear or phosphor-coated types, both considered to have better color rendition than MV. The MH lamp is used for practically every type of interior and exterior lighting because the MH possesses what is thought to be the most efficient "white" light source available. The efficiency of MH lamps is from 1.5 to 2 times that of MV lamps. MH lamp sizes range from 175 to 1500 W. Ballasts designed specifically for MH lamps must be used.

High-Pressure Sodium (HPS) Lamps.   Production of light within an HPS lamp is basically similar to that in a MV lamp. HPS lamps produce light when electrical current passes through a sodium vapor. HPS lamps have two envelopes, the inner one being made of a polycrystalline alumina in which the light-producing arc is struck. The outer envelope is protective, and may either be clear or coated. Because the sodium in the lamp is pressurized, the light produced is not the characteristic bright yellow associated with sodium, but rather a "golden white" light. There is no mercury radiation (ultraviolet) in the light output.

Although the HPS lamp first found its principal use in street and outdoor lighting, HPS is now a readily accepted light source in industrial plants, and is also being used in many commercial and institutional applications as well. HPS lamps have the highest lamp efficiency of all lamps normally used indoors.

Starting characteristics are similar to those for MV lamps, except that the starting time for HPS is approximately 15 to 20 minutes. HPS lamp sizes range from 70 to 1000 W. Ballasts designed specifically for HPS lamps must be used. The typical life rating of the HPS lamp is 24,000 hours. This, plus excellent *lumen maintenance* (Sec. 7-3), give the HPS lighting system a clear-cut economic advantage over an MV, MH, or fluorescent system.

Low-Pressure Sodium (LPS) Lamps.   LPS lamps are similar to HPS lamps in operating characteristics, except that the output of the LPS lamp is concentrated almost exclusively in a narrow wavelength (the yellow region). Thus colors other than yellow that are present in an object appear distorted under the LPS source. LPS lamps produce a monochromatic light output (reds, blues, and other colors illuminated by an LPS light source all appear as tones of gray). However, LPS lamps can be used where color discrimination is not an essential factor, but where high efficiency, low wattage, and other factors influencing system cost, installation, and maintenance are of real and continued value. The primary use of the LPS lamp is in street and highway lighting, outdoor area and security lighting (pedestrian crossings, tunnels, underpasses, parking lots), and in those indoor applications such as warehouses where color is not important.

7-3 CALCULATING THE LIGHTING REQUIREMENTS

The accepted procedure for calculating interior lighting levels is by use of the *zonal cavity system* of the Illuminating Engineering Society (IES). Although these IES calculations provide the best lighting design, and are fully described in the IES handbook, it is often necessary to have only a quick approximation of the quantity of lighting equipment needed to satisfy an illumination-level specification. There are several rules of thumb which serve that purpose. Before we get into these simplified methods, let us consider some thoughts on planning and choosing commercial and industrial lighting systems. Again, the General Electric Company line of equipment is used in our examples.

Most commercial and industrial functions can be lighted with any of the common lamp–luminaire combinations. However, when you want the most economy and best energy management, you must select the *most efficient system compatible with the task requirement*. Planning and choosing a lighting system requires knowledge of lamp and luminaire performance, as well as understanding the task.

7-3.1 Lighting According to the Task

Figure 7–5 shows recommended illumination levels (in footcandles) for various tasks. These recommendations are found in the American National Standard Practice for Office Lighting, sponsored and published by the IES of North America. Although the recommendations apply primarily to office locations, the illumination levels are the same for any similar commercial/industrial lighting system.

In applying the recommended levels of illumination shown in Fig. 7–5, remember that the level is *minimum* for the task. With uniform lighting, illumination falls off near walls and in the corners. Proper fixture location minimizes this effect. Light level also diminishes with gradual dirt accumulation on fixtures and lamps, as well as with normal lamp depreciation. Such depreciation is minimized with proper lamp and luminaire maintenance. The IES zonal cavity system takes the factors into account. If you are involved primarily in lighting systems, it is essential that you understand the zonal cavity system, and you will do well to study the IES handbooks and literature.

Area and task	Footcandles
Offices:	
Drafting rooms	
Detailed drafting and designing, cartography	200[b]
Rough layout drafting	150[b]
Accounting offices: auditing, tabulation, bookkeeping, business machine operation, computer operation	150[b]
General offices	
Reading poor reproductions, business machine operation, computer operation	150[b]
Reading handwriting in hard pencil or on poor paper, reading fair reproductions, active filing, mail sorting	100[b]
Reading handwriting in ink or medium pencil on good-quality paper, intermittent filing	70[b]
Private offices	
Reading poor reproductions, business machine operation	150[b]
Reading handwriting in hard pencil or on poor paper, reading fair reproductions	100[b]
Reading handwriting in ink or medium pencil on good-quality paper	70[b]
Reading high-contrast or well-printed materials	30[b]
Conferring and interviewing	30[b]
Conference rooms	
Critical seeing tasks	100[b]
Conferring	30[c]
Note-taking during projection (variable)	30[b, d]
Washrooms	30[c]
Elevators, escalators, stairways	20[c]
Corridors	20[d, e]

[a] Equivalent sphere illumination (ESI): the level of sphere illumination which would produce task visibility equivalent to that produced by a specific lighting environment.

[b] Minimum on the task at any time for young adults with normal and better than 20/30 corrected vision.

[c] Footcandles as measured with a light meter (rather than ESI).

[d] Controllable (dimmer).

[e] But no less than 20% of illumination in adjacent areas.

FIGURE 7-5   Recommended illumination levels (in footcandles) for various tasks.

7-3.2   Choosing the Lamp

As discussed (from a lighting standpoint) there are three major lamp types: incandescent, fluorescent, and HID. Within the HID category, there are three classifications in the General Electric Company line: Lucalox®, Multi-Vapor®, and mercury vapor. All types differ in electrical characteristics, efficacy, efficiency, lumen maintenance, life, color, size, shape, and cost. (In lighting terminology "efficacy" refers to a lamp's lumens per watt; "efficiency" refers to the fixture's lumen output divided by the lamp's lumen output.)

Figure 7–6 shows the approximate range of lamp characteristics for the General Electric Company line. The following paragraphs describe these characteristics.

Electrical Characteristics. The incandescent lamp operates directly from the voltage for which it is designed. All other lamp types require auxiliary equipment which transforms and/or regulates voltage, limits current, and so on. From a wiring design standpoint, *these auxiliary devices consume wattage,* in addition to the wattage consumed by the lamp.

Incandescent and fluorescent lamps attain full output as soon as they are energized. HID lamps require a warm-up time. In the event of even a momentary power interruption, HID lamps must cool for a period before their arcs restrike and light output is resumed. Both the warm-up and restrike times vary as shown in Fig. 7–6.

Efficacy. A lamp's lumen output is rated (as shown in the GE lamp catalog) according to average initial lumens. Dividing that value by lamp wattage determines the efficacy in lumens per watt (LPW). The LPW varies considerably among the several types of lamps, and also varies within the lamp types (the higher wattages usually produce more LPW than the lower wattages). All other factors being equal, the most economical lighting system is normally the one that uses lamps having the highest efficacy. Figure 7–7 shows an LPW comparison for General Electric Company lamps. Note that the ballast wattage is not included in Fig. 7–7.

Lumen Maintenance. Efficacy depreciates as a lamp is used. Again, the amount (and rate) of reduced LPW varies with lamp types and wattage. It is necessary to be familiar with lumen maintenance (the opposite of "depreciation") because illumination levels are normally calculated as *maintained footcandles* (a level which can be expected after the lamps have been operated for some specified time). The GE lamp catalog provides "mean" lumen ratings for both fluorescent and HID lamps.

Lamp Life. The rated life of a lamp (another item included in the GE lamp catalog) is that number of hours at which it is expected that 50% of a group of lamps is likely to fail. Lamp life is an important factor in lighting maintenance (replacement and labor) costs. However, long life, by itself, does not necessarily contribute to low overall cost of a lighting system.

Color. Different lamp types generate light which is composed of energy from various portions of the visible spectrum. Incandescent and Lucalox lamps appear "warm"; mercury vapor, fluorescent, and Multi-Vapor appear "cool." All of these lamp types produce completely acceptable illumination but, be-

Type of lamp	Wattages[a]	Average lumens per watt		Percent lumen maintenance	Rated average life (hours)	Warm-up/restrike (minutes)	Relative cost of light
		Initial	Mean				
Lucalox	35–1000	64–140	58–127	90–91	20,000–24,000[b]	3–4/0.5–1	Lowest
Multi-Vapor	175–1000	80–115	62–92	77–80	10,000–20,000	2–5/10–20	Medium
Mercury vapor (deluxe white)	50–1000	32–63	25–48	75–89	16,000–24,000+	5–7/3–6	High
Fluorescent	34–215	74–100	49–92	66–92	12,000–20,000+	Immediate	Medium
Incandescent	100–1500	17–24	15–23	90–95	750–2500	Immediate	Highest

[a] Lucalox: 10 sizes; Multi-Vapor: 6 sizes; mercury vapor: 7 sizes; fluorescent and incandescent: range most commonly used for industrial applications.

[b] 16,000 for 35-W Lucalox; 10,000 hours for 250-W Delux Lucalox.

FIGURE 7-6 Approximate range of lamp characteristics.

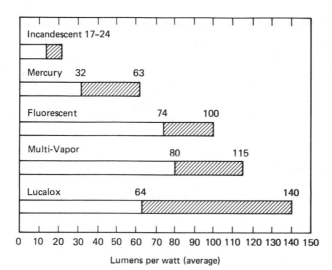

FIGURE 7-7 Comparison of lamps on the basis of lumens per watt (ballast watts not included).

cause of their spectral characteristics, tend to emphasize colors in which they are strong, and demean those colors in which the lamps are weak.

Cost. Unit lamp cost is a small part of lighting system cost. Although lamps with higher light output usually cost more, a smaller quantity is needed for a specified lighting level. Lighting costs tend to decrease when higher-priced lamps are used.

Lucalox high-pressure sodium lamps are usually the best economic choice for general industrial lighting. Lucalox lamps are high in efficacy (more than twice that of mercury), lumen maintenance, and life. These combined factors more than offset the higher cost of Lucalox.

Multi-Vapor metal-halide lamps, next to Lucalox in efficacy, are usually the second economic choice. While providing better color rendition and cooler appearance, the lower initial light output and poor lumen maintenance usually results in a higher cost of light. Also, the selection of different wattages for Multi-Vapor lamps is more limited than other HID lamps.

Mercury vapor lamps, successfully used by the millions for many years, are gradually being displaced by Lucalox and Multi-Vapor lamps. Mercury lamps are now generally used for certain low-footcandle, low-mounting-height applications, usually out of doors.

Fluorescent lamps (while slightly lower in efficacy than Multi-Vapor) may be preferred in locations where, by virtue of large surface area and diffusion, fluorescent produces softer reflections in shiny materials and/or equipment. Battery-operated fluorescent fixtures are useful for providing emergency lighting for safety. Where excellent color rendition is required, specially designed fluorescent lamps must be used.

Incandescent lamps are seldom used for general industrial lighting because of their low efficacy and comparatively short life. In commercial applications, incandescent lamps are generally being replaced by fluorescents. Both incandescent and fluorescent lamps are useful in seldom-occupied areas and/or where *instant turn-on is required.* Quartzline® incandescents are General Electric special-purpose lamps. Like conventional incandescents, Quartzline lamps provide instant on–off operation, with no warm-up required. Quartzline lamps are somewhat more efficient, maintain maximum light output roughly twice as long, and produce a whiter light. In industrial applications, Quartzline lamps are used primarily for supplemental lighting.

7-3.3 Choosing the Luminaire (Fixture)

Commercial and industrial luminaires or fixtures are so varied that it may be difficult to make the proper choice. Other than their design for specific types of lamps, the differences in fixtures lie in the *photometric, mechanical,* and *electrical* characteristics.

Photometric Properties of Luminaires.   The photometric properties of a luminaire are those that determine how the light is controlled and delivered to the functional area. Photometric information includes light distribution, efficiency, coefficient of utilization (CU), shielding, and brightness.

Candlepower Distribution.   Although luminaire distribution is classified by six types (by the IES), the *direct* and *semidirect* are used for most industrial lighting. The semidirect luminaires are generally preferred because they permit a percentage of light (10 to 40%) to be directed toward the ceiling (to reduce brightness contrast). Upward-component fixtures tend to be more efficient and stay cleaner due to the ventilation action of air moving through the fixture.

Beam Spread.   The downward-component of a luminaire's light distribution is further classified according to the beam spread, ranging from *highly concentrating* to *widespread.* In turn, these classifications are associated with a range of maximum spacing-to-mounting height (S/MH) ratios. Figure 7–8

Luminaire classification	S/MH above work plane
Highly concentrating	Up to 0.5
Concentrating	0.5 to 0.7
Medium spread	0.7 to 1.0
Spread	1.0 to 1.5
Wide spread	Over 1.5

FIGURE 7-8   Definition of S/MH ratios.

defines the S/MH ratios. The following is an example of how Fig. 7–8 is to be used.

If the bottom of the fixtures are 33 ft above the floor, and the work plane is 3 ft above the floor, the effective mounting height is 30 ft. Maximum spacing for a fixture with an S/MH of 0.6 is 18 ft (30 × 0.6). Never exceed S/MH ratios. Fewer shadows and more even lighting result if the fixtures in our example are on 12-ft rather than 18-ft centers.

Within any functional area, uniform horizontal illumination is obtained by selecting fixtures that provide adequate overlapping of the beams (on the installed spacing). Low mounting heights often require broad beam spreads; high mounting heights can often use narrow beam spreads.

Very few commercial and industrial tasks have a purely horizontal orientation; lighting is also needed on the vertical components. Also, optimum visual comfort requires that the vertical surface of the room and equipment are illuminated. Fixtures with excessive concentrations and insufficient beam overlapping can cause defective vertical lighting and harsh shadows. Wider beam spread provides better overlapping and a greater vertical component.

Efficiency. As discussed, efficiency is the ratio of a fixture's output to the lamp's output. Unfortunately, highly efficient fixtures may be lacking in adequate shielding or may misdirect the light. So it is not necessarily wise to select a fixture on efficiency alone. However, the efficiency rating serves to compare fixtures, particularly where other performance characteristics are similar.

Coefficient of Utilization (CU). A luminaire does not deliver all of the light to the work plane. The quantity is reduced by absorption within the fixture and the room. The ratio of useful light that reaches the work plane to the light output of the lamp is called the CU.

Shielding and Reflectors. Visual comfort is increased as the *shielding angle* (shown in Fig. 7–9) prevents light from hitting the eyes. The reflector also controls *direct glare,* or excessive brightness coming directly from the lamp or areas of the reflector. (It is not necessary to look directly into a fixture to experience glare; such annoyance may be apparent even in the peripheral vision). The shielding characteristic of a luminaire depends on the shielding

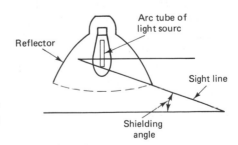

FIGURE 7–9 Relationship of luminaire reflectors and shielding angle.

angle shown in Fig. 7–9. As shown, the shielding angle is that angle between a horizontal line and the line-of-sight at which the source first becomes visible. Luminaire shielding is provided by the reflector edges and/or by auxiliary devices such as louvers, baffles, or lenses.

In factory areas, shielding should be at least 25°, preferably approaching 45° (a figure frequently used for office lighting). Fixtures with increased shielding should be used for higher illumination levels and/or lamps with high brightness.

Luminaire Brightness. Brightness is a function of lamp brightness, reflector contour, the material from which the reflector is made, and the shielding angle. Usually measured in footlamberts, the brightness information is found in manufacturers' photometric tables and should be used to compare fixtures.

Mechanical Features. A well-made fixture is easier to install and to maintain. Some of the important mechanical aspects to evaluate are material, construction, and auxiliary devices.

Good-quality industrial fixtures are somewhat more rugged than office fixtures. Reflectors for fluorescents are finished with baked, synthetic enamel, or with porcelain enamel. Although industrial fluorescent luminaires are not normally equipped with lateral baffles, such devices are frequently recommended because they provide optimum brightness control, irrespective of occupant–fixture orientation.

HID luminaires are being used more frequently in many applications. Most HID reflectors are made of prismatic glass, aluminum, or a glass–aluminum combination. Another HID luminaire is the enclosed type, which prevents dirt from coming in contact with the reflecting surface. A flat, gasketed, clear-glass cover seals the bottom; a clear-glass, gasketed collar seals the top. The enclosed luminaire can be vented to permit air to exit when the lamp is turned on, and to enter when the lamp is off. A filter (usually ceramic or activated charcoal) keeps dust and dirt out, and keeps the reflector clean and efficient.

7–3.4   Guidelines for Estimates in Lighting Design

When all of the information described in the IES Lighting Handbook for the zonal cavity system is not available, it is often necessary to make an educated guess. In fact, it is often possible to produce a rough (but sufficiently accurate) estimate of lighting requirements if a few simple guidelines are followed. This section describes guidelines sufficient for an electrician or electrical contractor to determine proper lighting requirements for commercial and industrial applications. The guidelines here are based on the tables shown in Figs. 7–10 and 7–11, which are used by the General Electric Company for quick estimates of lighting requirements.

Lamp	Watts per square foot for 100 footcandles
Lucalox	1.6
Multi-Vapor	2.5
Fluorescent	3.0
Mercury vapor	3.5
Incandescent (reflector lamp)	8.5

FIGURE 7-10 Watts-per-square-foot method of calculating lighting requirements.

Lamp	Watts	Approximate footcandles on spacings of:				
		10 ft X 10 ft	15 ft X 15 ft	20 ft X 20 ft	25 ft X 25 ft	30 ft X 30 ft
Lucalox	70	35	15	10	—	—
	100	55	25	15	10	—
	150	95	45	25	15	10
	250	180	80	45	30	20
	400	300	135	75	50	35
	1000	—	—	210	135	95
Multi-Vapor	175	85	35	20	15	10
	400	200	90	50	35	25
	1000	—	300	165	105	75
Mercury vapor	50	10	—	—	—	—
(deluxe white)	75	15	—	—	—	—
	100	20	10	—	—	—
	175	45	20	10	—	—
	250	60	30	15	10	—
	400	115	50	30	20	15
	1000	300	140	80	50	35

Fluorescent (cool white)	Continuous rows of two-lamp fixtures on spacings of:				
	6 ft	8 ft	10 ft	12 ft	15 ft
40-W rapid start	120	90	70	60	50
75-W slimline	120	90	70	60	50
110-W high output	185	140	110	90	75
215-W power groove	300	225	180	150	120

FIGURE 7-11 Lamp spacing method of calculating lighting requirements.

Watts per Square Foot. The table in Fig. 7–10 shows the approximate power, in watts, required to provide 100 footcandles of light for each square foot of area, using various types of lamps. Note that the Lucalox lamp requires the least power, while the incandescent lamp (with reflector) requires the most power. This is because the Lucalox lamp has the highest lumens-per-watt rating, while the incandescent has the lowest rating.

Using the information of Fig. 7–10 alone, it is possible to find the total wattage required for a given area. With the wattage established, the lighting branch circuit can then be designed. Conductor size, overcurrent and short-circuit protection, and contactor ratings can be calculated as described in Chapters 1 through 6 and Sec. 7–4.

Lamp Spacing. The table of Fig. 7–11 shows the approximate amount of light (in footcandles) that is available when various types of lamps (of various wattages) are used on given spacings.

In the case of fluorescents, the footcandle ratings of Fig. 7–11 are based on continuous rows of two-lamp fixtures. In using these fluorescent ratings, it is assumed that (1) the spacings are within the maximums established by the fixture manufacturer, (2) each fixture contains two lamps, and (3) the rows are continuous (covering either the full length or full width of the area). For example, the 40–W rapid-start fluorescent is 4 ft long. Generally, the nominal lamp length given in catalogs is the overall dimension of the lamp, including the lamp holders in which the lamp is seated. However, there are exceptions. Since there are two lamps per fixture, each 4 ft of the row uses 80 W (plus the ballast wattage), which brings the power to about 96 W per each 4 ft of row). If the rows are spaced 6 ft apart, the light is 120 fc.

Relationship between Spacing and Watts. Now let us see the relationship between Figs. 7–10 and 7–11. Assume that you want to light a hall-like area 100 ft long by 12 ft wide (1200 ft²) using two rows of two-lamp fluorescent fixtures. Each fixture has two 40-W rapid-start fluorescents, and uses about 96 W of power, including ballast. The rows are 6 ft apart, with 3 ft between each row and the walls, as shown in Fig. 7–12. You want 100 fc of light throughout the area.

According to Fig. 7–10, you need 3600 W (if fluorescents are used) to produce 100 fc of light (1200-ft² area × 3 W = 3600 W). According to Fig. 7–11, the two rows of fluorescents with 6-ft spacing produce 120 fc, 20% more than needed. Also, with 50 fixtures (25 in each row, assuming that each lamp uses 4 ft including mounting fixture), the total power is 4800 W (50 fixtures × 96 W = 4800 W).

If cost or energy conservation is of particular importance, it is possible to reduce the number of lamps and fixtures by 20%. This will reduce the light to the 100 fc level. The 6-ft spacing between rows should be maintained, but the number of two-lamp fixtures can be reduced to 40 by increasing the space between fixtures.

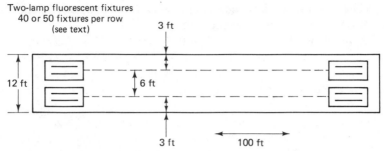

FIGURE 7-12 Lighting a hall-like area with two rows of two-lamp fluorescent fixtures.

HID Lamp Spacing.   In the case of HID lamps (Lucalox, Multi-Vapor, and mercury) the footcandle ratings of Fig. 7-11 are the *base-maintained footcandle levels,* according to fixture spacing. For footcandle levels other than the base quantity, the level is changed inversely, and proportionally, to a change in spacing. Doubling the spacing (in one direction) cuts the level to one-half (approximately). Cutting the spacing in half (in one direction) doubles the level (approximately). Doubling the spacing in both directions reduces the level to about one-fourth. For example, the footcandle rating of a 400-W Lucalox on 10 × 10 ft spacings is 300 fc. If the spacing is doubled in both directions to 20 × 20, the footcandle rating is reduced to 75 fc (exactly one-fourth of 300 in this case).

Area Shapes and Fixture Efficiency.   There are many variables in lighting design, so you must use the information in Fig. 7-10 and 7-11 as guidelines only. Also, you can apply correction factors to the information. The following are two examples.

The ratings of Fig. 7-11, both fluorescent and HID, are based on large industrial areas where room width (W) is six times the fixture mounting height (MH), or W = 6MH. If the ratio is changed so that W = 3MH, you must reduce the footcandles and increase the watts by 15%. That is, you must increase the wattage by 15% to get the same footcandles. Or, for a given wattage, you must reduce the footcandle rating by 15%. If the ratio is changed so that W = MH, reduce the footcandles 50%, or increase the wattage 50%.

Lucalox fixtures and incandescent reflector lamps tend to be more efficient and have higher lumen-maintenance characteristics. For either Lucalox or incandescent reflector lamps, increase the footcandles by 25%, or decrease wattage by 25%.

Example of Simplified Lighting Design.   Now let us use the information in Figs. 7-10 and 7-11 to make an estimate of the lighting requirements for a typical industrial application. Assume that you want to light a machine shop 65 ft wide by 160 ft long, with a mounting height of 60 ft. You want about

100 fc of light at the work plane. You also want maximum efficiency (energy conservation), so you select Lucalox lamps.

The machine shop area is 10,400 ft² (160 × 65). According to Fig. 7-10, you need about 16,640 W to produce 100 fc of light (10,400 × 1.6). You can get this with 16 1000-W Lucalox lamps. (Since the higher-wattage lamps of a given type produce the greatest efficiency, always start with the largest available lamp size. Then, if some other practical consideration prevents using the largest size, work down to the next largest size.)

It is also possible to get the desired amount of light with less wattage, by proper spacing. The ratio of width to length in our machine shop is not square, so the square spacings of Fig. 7-11 cannot be used directly. However, with spacings of 20 × 30 ft, in a pattern similar to that of Fig. 7-13, you can

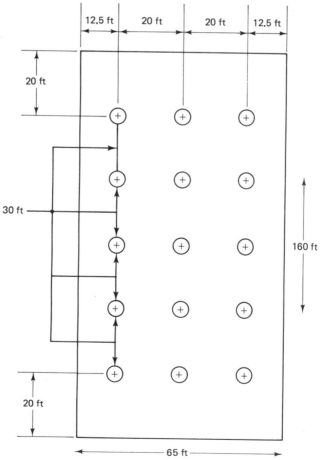

FIGURE 7-13   Lighting a machine shop with 15 1000-W Lucalox lamps.

use 15 1000-W Lucalox lamps. Of course, you must reduce the 20×20 rating of 210 fc shown for 1000-W Lucalox lamps in Fig. 7–11, since you are increasing the spacing in one direction (20×20 to 20×30). This reduces the 210 fc to about 160 fc.

Also, since the mounting height is almost the same as width, you should reduce the values of Fig. 7–11 by 50%, which brings the 160 fc down to 80 fc. However, the increase in efficiency of Lucalox fixtures offsets this factor by about 25%, and brings the 80 fc back up to 100 fc.

Lighting Branch-Circuit Currents. We have calculated the required number of lamps, spacing, and wattage for the machine shop lighting, using simple guidelines. Now we must determine the actual current consumed by these 15 1000-W Lucalox lamps so that we can design the lighting branch circuit and select a suitable lighting contactor.

You cannot simply divide the power by voltage, since Lucalox are HID lamps, and use a ballast. According to the GE Lamp Catalog, 1000-W Lucalox lamps use ANSI Specification S52 ballasts. According to the GE ballast and energy systems catalog, the S52 ballast can be used with voltages of 120/208/240/277/480 V, and require corresponding input currents of 9.1/5.1/4.6/4.0/2.3 A. As a practical matter, the choice of voltage probably depends on what source is readily available. If 120 V is used, each lamp (including ballast) draws 9.1 A, and the 15 lamps draw 136.5 A. The nearest standard circuit breaker is 150 A. The nearest standard lighting contactor is 200 A. We discuss contactors next.

7-4 LIGHTING CONTACTORS

Most present-day commercial and industrial lighting systems require more than simple on–off control. Quite often, some form of remote control is required, and this control may or may not be in addition to a master control station at a central location. Certain applications include the use of automatic control by time clocks or photoelectric cells. Whatever the need, lighting contactors provide the most efficient means of controlling the system.

Typical areas that can be controlled efficiently by lighting contactors include parking lots, industrial plants, office buildings, theaters and auditoriums, hospitals and institutions, shopping centers, stadiums, and airports.

7-4.1 *Types of Lighting Contactor Control*

Lighting contactors should be part of any energy management system, since contactors help conserve energy and reduce utility bills by providing three types of control:

Lighting contactors offer both *centralized* and *remote control* of lighting. Circuits can be turned on and off from a number of remote locations, in addition to a master control station.

Lighting contactors also offer *selective switching* of lights. Selective switching is the control of one or more individual lighting circuits, independent of the other circuits. This design allows the potential for turning on only the amount of lighting that is actually needed.

Lighting contactors can provide *automatic control* to ensure that lights are turned off when not needed. There are a number of devices that, when used with lighting contactors, offer a convenient and reliable method of automatically controlling lighting loads. These devices include *program time clock, photoelectric cell, programmable controller,* and *demand controller.*

7-4.2   Centralized and Remote Control with Lighting Contactors

In many present-day lighting systems, the circuitry is designed so that the only method of controlling lighting is from a central panelboard. This is suitable when all of the areas require lighting at all times. However, during the period when only certain areas need illumination, all the lights must be on, because there is no way to energize only the few circuits that need to be on.

Lighting contactors offer both centralized and remote control of lighting, as shown in Fig. 7–14. Note that the lighting contactor types (Type S, Type L, etc.) shown in Fig. 7–14 are those manufactured by the Square D Company. Individual circuits can be turned on and off, independent of the other circuits, from a number of remote locations in addition to the master control station. This makes it easier for people to turn lights off when they leave the area, thus saving money on electrical bills. As shown in Fig. 7–14, the most commonly used methods for control of lighting contactors is with manual devices (such as toggle switches, pushbuttons, or selector switches) and automatic control devices (such as time clocks, photoelectric cells, or timing relays).

7-4.3   Electrically versus Mechanically Held Lighting Contactors

There are two basic types of lighting contactors: electrically held and mechanically held. Each has its own advantages and disadvantages.

Electrically Held.   All electrically held contactors require voltage to be *continuously applied* to the coil to maintain the contacts in the closed position (or open position in the case of normally closed contacts). Figure 7–15 shows the basic circuit of an electrically held lighting contactor. Operation of the circuit is as follows.

When the normally open ON switch is closed (by momentarily pushing

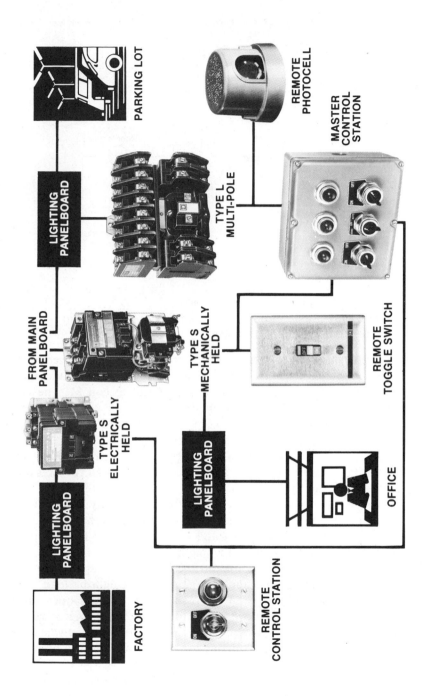

PARKING LOT

REMOTE PHOTOCELL

MASTER CONTROL STATION

LIGHTING PANELBOARD

TYPE L MULTI-POLE

FROM MAIN PANELBOARD

TYPE S MECHANICALLY HELD

REMOTE TOGGLE SWITCH

TYPE S ELECTRICALLY HELD

LIGHTING PANELBOARD

OFFICE

LIGHTING PANELBOARD

FACTORY

REMOTE CONTROL STATION

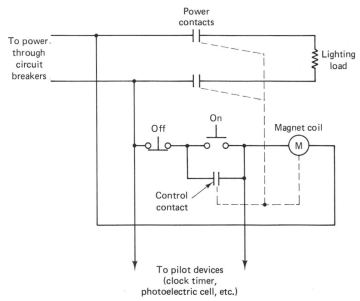

FIGURE 7-15 Basic circuit of an electrically held lighting contactor with common power supply connection.

the ON button, power is applied to magnetic coil M, causing all of the contacts actuated by coil M to close. The two power contacts apply power to the lighting load. The control contact closes and applies power to coil M. The ON button can then be released. The power applied through the control contact to coil M keeps all contacts closed and keeps power applied to the lighting load.

When the normally closed OFF switch is opened (by momentarily pushing the OFF button), power is removed from coil M, opening all contacts and removing power from the lighting load.

Note that the lighting contactor can also be operated by *pilot devices* such as a clock timer or photoelectric cell. The contacts within the pilot device replace those of the control circuit ON and OFF switches. When the pilot device contact closes (at a predetermined time in the case of a timer, or at a given light level in the case of a photoelectric cell), power is applied to coil M, closing the power contacts and applying power to the lighting load. The contacts remain in this condition until the pilot device contacts open (at a certain time or light level) and remove power to coil M, removing power to the lighting load. Note that the ON and OFF switches are not used when the pilot device is used. The contactor is under complete control of the pilot device.

Also note that the circuit of Fig. 7-15 is a *common power supply* connection. That is, the control circuit uses the same power as the lighting load. This is the most convenient arrangement. However, in those cases where the lighting load is operated at high voltages (such as 480 V), it is often desirable to operate the control circuit from a *separate control supply* of lower voltage (such as 120 V). Of course, this requires more elaborate wiring.

From a simplified wiring design standpoint, the main point to remember is that both the lighting contactor coil and the pilot device have voltage ratings that must be observed. Also, there are usually recommended wire sizes and lengths for conductors between the pilot devices and the lighting contactor coils. This subject is discussed further in Sec. 8-5.4.

Electrically held lighting contactors are used:

1. Wherever a high rate of operation is encountered (such as heating elements or electric furnaces).

2. In areas where the a-c hum of a continuously energized coil is not annoying (such as factories, machine shops, or outdoors).

3. With three-wire control schemes to provide low-voltage protection, thus preventing the load from being automatically energized after restoration of power following a failure.

Both two-wire and three-wire control are discussed in Chapter 8 (Sec. 8-5).

Mechanically Held.   All mechanically held contactors require only *momentary application* of voltage to be *latched* (turned on) or *unlatched* (turned off). Since the contacts are mechanically held closed (open if normally closed contacts), the latch and unlatch coils need only be momentarily energized, thus eliminating a-c hum. This feature allows for quiet operation of the mechanically held lighting contactor, and make the perfect choice for quiet locations. As standard Square D Company mechanically held lighting contactors are provided with *coil-clearing contacts,* assuring that the coils are deenergized even if the control device is held closed. Figure 7-16 shows the basic circuit of a mechanically held lighting contactor. Operation of the circuit is as follows.

When the normally open ON switch is closed (by momentarily pushing the ON button), the latch coil is energized, the power contacts are closed, and power is applied to the lighting load. Coil-clearing contacts A open (and contacts B close) once the power contacts are fully closed. This removes power to the latch coil (and makes power available to the unlatch coil). However, the control circuit is completely deenergized, eliminating any a-c hum.

When the normally open OFF switch is closed (by momentarily pushing the OFF button), power is applied to the unlatch coil (through the now-closed B contacts), opening the power contacts and removing power from the lighting load. Coil-clearing contacts B open (and contacts A close) once the power contacts are fully open.

Note that the mechanically held lighting contactor can also be operated by a pilot device (timer, photoelectric cell, etc.). However, the pilot device must be two-pole (one normally open contact and one normally closed contact). Also, the circuit of Fig. 7-16 uses the common power supply connection. A separate control supply can be used.

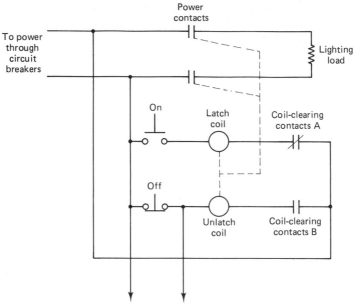

FIGURE 7-16   Basic circuit of a mechanically held lighting contactor with common power supply connection.

FIGURE 7-17   Type S electrically held lighting contactor. (Courtesy Square D Company.)

Mechanically held lighting contactors are used wherever:

1. Quiet operation is required (offices, hospitals, schools).
2. Circuits are required to remain closed (or open in the case of normally closed contacts) during a power failure.
3. Excessive control line distances are required.

7-4.4 Typical Lighting Contactors

The Square D Company manufactures two typical lighting contactors designated as Class 8903: the Type S lighting contactor and the Type L multipole lighting contactor. The following are brief descriptions of these devices.

Type S Lighting Contactor. The Type S lighting contactors are available in both electrically held (Fig. 7-17) and mechanically held (Fig. 7-18) versions, for lighting applications of 30 to 800 A. Type S contactors are rated

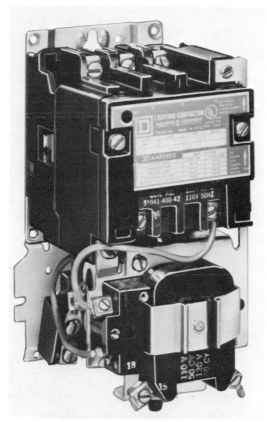

FIGURE 7-18 Type S mechanically held lighting contactor. (Courtesy Square D Company.)

Ampere rating	Number of poles	Enclosure type				
		NEMA Type 1	Flush mounting	NEMA Type 4	NEMA Type 12	Open type
		Electrically held				
30	2	SMG–1	SMF–1	SMW–1	SMA–1	SMO–1
	3	–2	–2	–2	–2	–2
	4	–3	–3	–3	–3	–3
	5	–4	–4	–4	–4	–4
60	2	SPG–1	SPF–1	SPW–1	SPA–1	SPO–1
	3	–2	–2	–2	–2	–2
	4	–3	–3	–3	–3	–3
	5	–4	–4	–4	–4	–4
100	2	SQG–1	SQF–1	SQW–1	SQA–1	SQO–1
	3	–2	–2	–2	–2	–2
	4	–3	–3	–3	–3	–3
	5	–4	–4	–4	–4	–4
200	2	SVG–1	SVF–1	SVW–1	SVA–1	SVO–1
	3	–2	–2	–2	–2	–2
	4	–3	–3	–3	–3	–3
300	2	SXG–1	—	SXW–1	SXA–1	SXO–1
	3	–2	—	–2	–2	–2
400	2	SYG–1	—	SYW–1	SYA–1	SYO–1
	3	–2	—	–2	–2	–2
600	2	SZG–1	—	SZW–1	SZA–1	SZO–1
	3	–2	—	–2	–2	–2
800	2	SJG–1	—	SJW–1	SJA–1	SJO–1
	3	–2	—	–2	–2	–2
		Mechanically held				
30	2	SMG–10	SMF–10	SMW–10	SMA–10	SMO–10
	3	–11	–11	–11	–11	–11
	4	–12	–12	–12	–12	–12
	5	–13	–13	–13	–13	–13
60	2	SPG–10	SPF–10	SPW–10	SPA–10	SPO–10
	3	–11	–11	–11	–11	–11
	4	–12	–12	–12	–12	–12
	5	–13	–13	–13	–13	–13
100	2	SQG–10	SQF–10	SQW–10	SQA–10	SQO–10
	3	–11	–11	–11	–11	–11
	4	–12	–12	–12	–12	–12
	5	–13	–13	–13	–13	–13
200	2	SVG–10	SVF–10	SVW–10	SVA–10	SVO–10
	3	–11	–11	–11	–11	–11
	4	–12	–12	–12	–12	–12
300	2	SXG–13	—	SXW–13	SXA–13	SXO–13
	3	–14	—	–14	–14	–14
400	2	SYG–16	—	SYW–16	SYA–16	SYO–16
	3	–17	—	–17	–17	–17
600	2	SZG–18	—	SZW–18	SZA–18	SZO–18
	3	–19	—	–19	–19	–19
800	2	SJG–10	—	SJW–10	SJA–10	SJO–10
	3	–11	—	–11	–11	–11

FIGURE 7-19 Current ratings, number of poles, and enclosure types for Type S lighting contactors.

for tungsten, fluorescent, and HID lighting loads, as well as being fully rated for resistance heating and mixed loads (which include motor, lighting, and heating loads). Type S lighting contactors use silver-alloy double-break contacts for longer life, enclosed arc chambers for added protection, and have UL-listed withstand ratings to 100,000 A.

Figure 7–19 shows the current ratings, number of poles, and enclosure types for Type S lighting contactors. Note that derating of Type S contactors is not necessary for lighting or heating loads (as is the case for some motor contactors described in Chapter 8). Also, Type S contactors are fully rated for motor loads and have horsepower ratings equal to NEMA-size motor contactors (which are discussed in Chapter 8).

Type L Multipole Lighting Contactor. The Type L multipole lighting contactors are available in both electrically held (Fig. 7–21) and mechanically held (Fig. 7–20) versions, for lighting applications where each circuit does not exceed 20 A. Type L contactors are rated for tungsten, fluorescent, and HID lighting loads, as well as being fully rated for resistance heating loads, with up to 12 convertible poles. However, Type L contactors are *not rated for mo-*

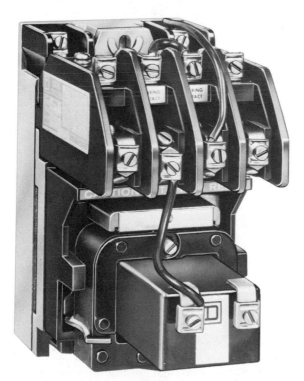

FIGURE 7–20 Type L multipole 20-A mechanically held lighting contactor. (Courtesy Square D Company.)

FIGURE 7-21   Type L multipole 20-A electrically held lighting contactor. (Courtesy Square D Company.)

Ampere rating	Number of poles	Enclosure type			
		NEMA Type 1	NEMA Type 4	NEMA Type 12	Open type
		Electrically-held			
20	2	LG–20	LW–20	LA–20	LO–20
	3	–30	–30	–30	–30
	4	–40	–40	–40	–40
	6	–60	–60	–60	–60
	8	–80	–80	–80	–80
	10	–1000	–1000	–1000	–1000
	12	–1200	–1200	–1200	–1200
		Mechanically-held			
20	2	LLG–20	LLW–20	LLA–20	LLO–20
	3	–30	–30	–30	–30
	4	–40	–40	–40	–40
	6	–60	–60	–60	–60
	8	–80	–80	–80	–80
	10	–1000	–1000	–1000	–1000

FIGURE 7-22   Current ratings, number of poles, and enclosure types of Type L lighting contactors.

tor loads. Type L contactors use silver-alloy double-break contacts, which are convertible from normally open (NO) to normally closed (NC), and are replaceable.

Figure 7–22 shows the current ratings, number of poles, and enclosure types for Type L contactors. For high-fault current protection, the Type L is the only multipole lighting contactor available with a UL-listed short-circuit withstand rating up to 22,000 A at 240 V.

7-4.5 *Type S Combination Lighting Contactors*

It is often desirable to install a disconnect means, overcurrent protection, and a lighting contactor in one enclosure. Figures 7–23 and 7–24 show such combination lighting contactors. The combination unit of Fig. 7–23 has a heavy-duty visible-blade switch, fusible or nonfusible, and features box-lug line connections. The unit in Fig. 7–24 has thermal-magnetic circuit breakers with

FIGURE 7-23   Type S combination lighting contactor with disconnect switch. (Courtesy Square D Company.)

FIGURE 7-24   Type S combination lighting contactor with circuit breaker. (Courtesy Square D Company.)

push-trip features for easy operational check. The circuit breakers provide short-circuit protection and can easily be reset after tripping. Alarm switches, undervoltage trip, and auxiliary interlocks are also available.

7-4.6   Outdoor Lighting Controllers

Square D Class 8903 Night-Master® outdoor lighting controllers offer a disconnect means, overcurrent protection, and a lighting contactor in one enclosure designed specifically for outdoor lighting. These combination units satisfy the requirements of both NEC and UL for service entrance equipment. Figure 7-25 shows a typical unit, which offers an advantage over enclosed safety switches or circuit breakers in that the electrically held contactor can provide automatic or remote control for all types of lighting. When used with photoelectric cells, time clocks, or other control devices, the controllers offer an economical means of controlling outdoor lighting. Typical applications include control of lighting for streets, parking lots, billboards, tennis courts,

FIGURE 7-25 Typical outdoor lighting controller. (Courtesy Square D Company.)

stadiums, country clubs, and parks. The controllers are available in both short and long versions, 30 to 200 A, with either fusible disconnect switch or circuit breaker. The enclosure is a NEMA Type 3 R, rainproof for outdoor applications, and has provisions for adding control units, solid neutral, and ground lugs.

7-4.7 Installation of Lighting Contactors

For new installations, lighting contactors can be installed either in the lighting panelboard, as shown in Fig. 7-26, or in their own enclosure next to or remote from the panelboard. In existing applications where the lighting control system is being updated, lighting contactors can be installed in their own enclosure next to a lighting panelboard as shown in Fig. 7-27.

7-4.8 Withstand Ratings for Lighting Contactors

Lighting contactors have short-circuit withstand ratings similar to those of switches and circuit breakers (Chapter 3). As in the case of any contacting or

FIGURE 7-26 New installation where lighting contactor is installed in the lighting panelboard. (Courtesy Square D Company.)

controlling device, *the short-circuit current available at lighting contactors must not exceed the short-circuit withstand rating of the contactor.* Square D Type S lighting contactors, with continuous current ratings of 30 through 400 A, are available with short-circuit withstand ratings up to 100,000 A, while the Type L lighting contactors have withstand ratings up to 22,000 A.

Figure 7-28 shows the withstand ratings for lighting contactors used with fusible disconnect switches. Note that Class R fuse clips are required, that only Class RK5 fuses are listed, and that the ratings apply to Type S contactors (since Type L contactors have a limit of 20 A of continuous current).

Figure 7-29 shows the withstand ratings for lighting contactors used with circuit breakers. The ratings of Fig. 7-29 apply to both Type S and Type L contactors.

The withstand ratings of both Figs. 7-28 and 7-29 apply to contactors in all enclosure configurations and are in rms symmetrical amperes.

FIGURE 7-27 Existing installation where the lighting contactor is installed in an enclosure next to the lighting contactor. (Courtesy Square D Company.)

7-4.9 *Continuous Lighting-Load Current Ratings for Lighting Contactors*

Each set of contacts in the Type L lighting contactor is rated at a maximum of 20 A for a lighting load, as shown in Fig. 7–22. The ratings for Type S lighting contactors (for lighting loads) are given in Fig. 7–19. Both Type L

Contactor continuous rating (A)	Maximum Class RK5 fuse rating (A)	Maximum voltage	Withstand rating (A)
30	30	600	100,000
60	60	600	100,000
100	100	600	100,000
200	200	480	100,000
300	400	600	100,000
400	400	600	100,000

FIGURE 7-28 Withstand, fuse, and voltage ratings for lighting contactors (Type S) used with fusible disconnect switches.

Contactor continuous rating (A)	Maximum circuit breaker rating (A)	Recommended Square D circuit breaker types	Maximum voltage	Withstand rating (A)
20	25	EH-EHB, FH-FHL	240	22,000
20	25	EH-EHB, FA-FAL, FH-FHL	480	14,000
30	40	FA-FAL, FH-FHL	600	10,000
30	40	IF-IFL	480	100,000
60	80	FH-FHL	600	18,000
60	90	IF-IFL	480	100,000
100	125	KA-KAL, KH-KHL	600	10,000
100	125	IK-IKL	480	100,000
200	250	LA-LAL	600	14,000
200	225	IK-IKL	480	100,000
300	400	LA-LAL, LH-LHL	600	22,000
400	800	MA-MAL, MH-MHL	600	22,000
600	800	MA-MAL, MH-MHL	600	22,000

FIGURE 7-29 Circuit breaker, voltage, and withstand ratings for lighting contactors (Type S and L) used with circuit breakers.

and Type S contactors are *fully rated* for tungsten, ballast, and resistance loads. This means that either type of contactor can be used to control such loads up to the full nameplate rating. Derating of the contactor is not necessary, as is standard practice with circuit breakers and fuses. *The NEC requires that the continuous load supplied by a branch circuit not exceed 80% of branch circuit rating.* For example, the load must not exceed 20 A in a branch circuit with 25-A conductor and circuit breakers or fuses. However, the Type S or Type L lighting contactor need not be derated. A 20 A contactor is rated to handle a full 20-A lighting or resistance load.

Example of Lighting Contactor Selection. In Sec. 7–3, we calculated that the 15 Lucalox lamps used to light a machine shop require 136.5 A if 120 V is used. A Type L contactor cannot be used for this branch circuit since Type L is rated at a maximum of 20 A per circuit. The next largest Type S contactor is 200 A, as shown in Fig. 7–19. The next largest circuit breaker is 150 A (Sec. 3–2.2). This is well below the maximum circuit breaker rating for the 200-A contactor, as shown in Fig. 7–29.

7-4.10 Continuous Resistance-Load Current Ratings for Lighting Contactors

Square D lighting contactors are fully rated for resistance loads up to 600 V, and can be used on resistance-type boilers, electric furnaces, electric water heaters, and snow-melting cables and panels. The calculations for selecting a contactor of the proper rating are essentially the same as for lighting loads (Sec. 7–4.9). However, calculations for resistance loads are often given in kilowatts (kW), using three-phase power, since three-phase is so much more efficient where heavy currents are involved. Figure 7–30 can be used to simplify the process of selecting lighting contactors used for pure resistance loads.

Example of Selecting a Contactor for Resistance Loads. In Chapter 4, we calculated that a 30-kW resistance load, using 230-V three-phase, requires 76 A. The calculations involve dividing the 30 kW by 230 V, then multiplying the result by 1.732. The process can be simplified using Fig. 7–30. Simply find 230 V in the voltage column, and move across to a kilowatt rating higher than 30 kW, to find the correct contactor size (100 A).

7-4.11 Continuous Motor-Load Current Ratings for Lighting Contactors

As discussed in Chapter 8, motors have an inrush current or *locked-rotor current* of approximately six times the full-load current. Square D Type S lighting contactors are fully rated for motor load, and have *horsepower* (hp) ratings equal to equivalent NEMA-size motor contactors (Chapter 8), as shown in Fig. 7–31. As discussed in Chapter 8, NEMA motor contactor sizes run from 00, 0, and 1 through 8, so the only size motor contactor that cannot be replaced by a Type S lighting contactor is NEMA size 8. (This is for 450-hp motors operating at 230 V and 900-hp motors operating at 460 or 575 V.)

Three-phase voltage	Lighting contactor size							
	30 A	60 A	100 A	200 A	300 A	400 A	600 A	800 A
200	10.3	20.7	34.6	69.2	103.9	138.5	207.8	277.1
230	11.9	23.9	39.8	79.6	119.5	159.3	239.0	318.7
380	19.7	39.4	65.8	131.6	197.4	263.2	394.9	526.5
460	23.9	47.8	79.6	159.3	239.0	318.6	478.0	637.4
575	30.0	60.0	99.0	199.0	299.0	398.4	597.6	796.7

FIGURE 7-30 Continuous resistance-load current ratings for lighting contactors.

Lighting contactor size (A)	Horsepower rating same as equivalent NEMA-size motor contactor
30	NEMA size 1
60	NEMA size 2
100	NEMA size 3
200	NEMA size 4
300	NEMA size 5
400	—
600	NEMA size 6
800	NEMA size 7

FIGURE 7–31 Continuous motor-load current ratings for lighting contactors (corresponding to NEMA-size motor contactors).

7–4.12 *Continuous Mixed-Load Current Ratings for Lighting Contactors*

All Type S lighting contactors are rated to handle mixed loads, or a combination of lighting, resistance, and motor loads. Figure 7–32 makes the selection of lighting contactors for mixed loads easier. Simply follow these steps:

1. Determine the total load by summing the lighting, resistance, and motor (full-load currents of each motor) loads.

2. Calculate the lighting and resistance (total nonmotor) load as a percentage of the total load.

3. From that percent column in Fig. 7–32, determine the size of lighting contactor required at the specified motor voltage.

As an example, assume that a lighting contactor is to control a mixed load where the total motor current is 90 A and the total lighting/resistance load is 80 A, using 200-V three-phase. This represents a combined load of 170 A, so a 200-A contactor is required. The lighting/resistance load is greater than 25%, but less than 50% of the total, so the 50% column should be used. The intersection of the 200-V three-phase, 200 A, and 50% columns shows that the maximum nonmotor current up to 100 A, and a 30-hp motor, can be controlled by 200-A lighting contactor.

Figure 7–32 can also be used to find the relationship between lighting contactor current ratings and motor horsepower ratings. Simply use the 0% column. For example, assume that a 450-hp motor must be controlled by a lighting contactor using 380 V three-phase. No other loads are connected to the branch circuit. Figure 7–32 shows that an 800-A lighting contactor is required.

Motor voltage and phase	Contactor rating (A)	Percent of total load represented by lighting and/or resistance							
		0%		25%		50%		75%	
		Max. non-motor amps	Max. motor hp	Max. non-motor amps	Max. motor hp	Max. non-motor amps	Max. motor hp	Max. non-motor amps	Max. motor hp
200-V three-phase	30	0	7.5	7.5	5	15	3	22.5	1.5
	60	0	10	15	10	30	7.5	43	3
	100	0	25	25	20	50	15	75	5
	200	0	40	50	40	100	30	150	15
	300	0	75	75	75	150	50	225	20
	400	0	125	100	100	200	60	300	30
	600	0	150	150	150	300	100	450	50
	800	0	250	200	200	400	125	600	60
230-V three-phase	30	0	7.5	7.5	7.5	15	3	22.5	2
	60	0	15	15	10	30	10	45	3
	100	0	30	25	25	50	15	75	7.5
	200	0	50	50	50	100	30	150	15
	300	0	100	75	75	150	50	225	25
	400	0	150	100	100	200	75	300	30
	600	0	200	150	150	300	100	450	50
	800	0	300	200	200	400	150	600	75
380-V three-phase	30	0	10	7.5	7.5	15	7.5	22.5	3
	60	0	25	15	20	30	15	45	7.5
	100	0	50	25	40	50	30	75	10
	200	0	75	50	75	100	60	150	30
	300	0	150	75	150	150	100	225	40
	400	0	250	100	200	200	125	300	60
	600	0	300	150	250	300	150	450	75
	800	0	450	200	350	400	250	600	125
460–575-V three-phase	30	0	10	7.5	7.5	15	7.5	22.5	3
	60	0	25	15	20	30	20	45	10
	100	0	50	25	40	50	30	75	15
	200	0	100	50	100	100	75	150	30
	300	0	200	75	150	150	100	225	50
	400	0	300	100	200	200	150	300	75
	600	0	400	150	350	300	200	450	100
	800	0	600	200	500	400	300	600	150
115-V single-phase	30	0	2	7.5	1.5	15	0.75	22.5	0.5
	60	0	3	15	3	30	2	45	0.75
	100	0	7.5	25	5	50	3	75	2
230-V single-phase	30	0	3	7.5	2	15	2	22.5	0.75
	60	0	7.5	15	5	30	5	45	2
	100	0	15	25	15	50	10	75	3

FIGURE 7-32 Continuous mixed-load current ratings for lighting contactors.

Type of load	When connected, single pole to load	When connected, two poles to load on single-phase and three poles to load on three-phase	When connected, single pole to load	When connected, two poles to load on single-phase and three poles to load on three-phase
Type of load	Types L and LL: 20 A		Type SM: 30 A	
Tungsten	277	480	277	480
Ballast	277[a]	480[a]	347	600
Resistance	600	600	600	600
Control circuit (coil) voltage	12–600 Type L; 24–277 Type LL; 24, 32, 115/125, and 230/250 dc Type L (seven poles max.)		6–600	
Type of load	Types SP, SQ, SV, and SX: 60–300 A		Types SY, SZ, and SJ: 400–800 A	
Tungsten	277	480	—	—
Ballast	347	600	347	600
Resistance	600	600	600	600
Control circuit (coil) voltage	6–600 Type SP; 24–600 Types SQ and SV; 120–600 Type SX		120–600	

[a]Types L and LL contactors also have ballast lamp rating of 15 A, 347 ac, when connected one pole to load, and 600 ac when connected two poles to load on single-phase and three poles to load on three-phase.

FIGURE 7–33   Maximum a-c voltage ratings for lighting contactors.

Type of load	Types L and LL: 20 A	Type SM: 30 A	Types SP, SQ, SV, and SX: 60–300 A	Types SY, SZ, and SJ: 400–800 A
Dc with two poles in series	125	—	250	—
Dc with three poles in series	250	250	250	—

FIGURE 7–34   Maximum d-c voltage ratings for lighting contactors used with tungsten lamp or resistance load only.

7-4.13 Maximum Voltage Ratings for Lighting Contactors

When selecting lighting contactors, the maximum voltage rating of the device must be considered in addition to the current rating. Figure 7-33 lists the maximum a-c voltage ratings of both Types L and S lighting contactors for ballast, tungsten, and resistance loads. Figure 7-34 shows the applicable d-c voltage ratings.

Lighting contactors are considered as controllers, and a UL requirement is that the rating of controllers be clearly marked on the device. All Class 8903 Types L and S lighting contactors have the maximum current and voltage ratings printed on the nameplate, visible even after installation.

8

ELECTRIC MOTOR WIRING DESIGN AND CONTROL

From a mechanical design standpoint, an electric motor is considered to be a source of energy, and is usually rated in such terms as horsepower, foot-pounds of torque, speed, and so on. From an electrical design standpoint, an electric motor is simply another electrical load. That is, an electric motor requires a certain voltage and draws a certain amount of current. If there are several motors, and they all require the same voltage, they can all be connected in the distribution system in parallel.

Electric motors usually have some inductive reactance. So the load presented by an electric motor is an impedance, rather than a pure resistance, and a power factor is involved. Motors may be rated by voltage and current, or by voltage and wattage, with the power factor specified. Generally, motors have a lagging current since they are inductive. However, in certain circumstances, a synchronous motor draws a leading current. This makes it possible to use synchronous motors to correct a lagging power factor condition, or to bring a power factor within tolerance. This is discussed fully in Chapter 9.

Once the voltage and current for an electric motor (or group of motors) is established, the distribution system can be designed to provide the necessary power, using the information of Chapters 1 through 5 (conductor size, voltage drop, grounding, overcurrent protection, etc.). If different voltages are required for different motors in the system, transformers can be used, as discussed in Chapter 6.

In addition to designing the power distribution system to the motors, it is necessary to provide the motors with *control devices*. Even the simplest motors must have a means of starting and stopping, as well as overcurrent protection (in addition to that provided by the branch circuits of the distribution system). It is generally not necessary to design the motor control components (starters, controllers, contactors, etc.) because these are available as off-the-shelf items (as are the motors). However, it is necessary to understand operation of the motor control systems to provide the correct wiring and to select the correct motor control components.

The purpose of this chapter is to familiarize the electrician or contractor with motor control fundamentals. The basic theory of electric motors, starters, controllers, and relays is omitted. Instead, we concentrate on those electric motor characteristics that directly and indirectly affect the design of electrical wiring. For example, a certain class of starter is required for a motor of given voltage and current. This information is available from the starter manufacturer, usually in tabular form to simplify the design task. Also, the NEC has very specific requirements concerning motor branch circuit design. Such information is not duplicated in full here. Instead, we describe how the information is to be used in the wiring design, and how guidelines can be used to simplify design.

This chapter also provides definitions, symbols, diagrams, and illustrations that give the reader a sound background in the language and basic principles associated with motor control components. The material in this chapter is limited to a-c motors, operating at voltages less than 600 V. Direct-current and high-voltage motors have very special wiring design requirements.

8-1   ELECTRIC MOTOR BASICS

Before going into the practical design of electric motor wiring, let us review the basics of electric motors, the terms used, the most common types, and their characteristics.

8-1.1   The Squirrel-Cage Motor

The great majority of a-c motors in use today are of the squirrel-cage type. This motor gets its name from the rotor construction. The rotor has no wire windings. Rotor bars are used instead, and the rotor has ball bearings rather than sleeve bearings. As the rotor bars are cut by the stator flux, the bars have a voltage induced by transformer action. A current flows in the short-circuited rotor bars, causing a magnetic flux around the bars. This develops a torque, causing the rotor to follow the rotating field. Squirrel-cage motors are simple in construction and operation. Simply connect three power lines to the motor, and the motor runs.

8-1.2   *Motor Speed and Slip*

The speed of a squirrel-cage motor depends on the number or poles in the motor winding. With 60-Hz power, a two-pole motor runs at about 3450 r/min, a four-pole at 1725 r/min, and a six-pole at 1150 r/min. Motors are frequently referred to by their *synchronous speeds:* 3600, 1600, and 1200 r/min, respectively.

The difference between synchronous speeds and full-load speeds can be explained as follows. The starting current of a squirrel-cage motor is high but of very short duration, and decreases as the rotor approaches the speed of the rotating field. The rotor cannot reach this synchronous speed. If it could, no flux would cut the rotor, there would be no induction or rotor current or torque, and the motor would slow down. The actual rotor speed slips behind the rotating field sufficiently so that the enough current can be induced to produce the torque needed to satisfy the demands of the mechanical load.

The inability to keep up with the synchronous speed is an important measure of motor performance, and is called *slip* or *motor slip*. Slip may be measured in r/min or as a percentage of synchronous speed. When written as a decimal, the symbol for slip is *s* (hopefully not to be confused with seconds).

8-1.3   *Motor Torque and Full-Load Current (FLC)*

Torque is the ''turning'' or ''twisting'' force of the motor and is usually measured in pound-feet (or foot-pounds). Except when the motor is accelerating up to speed, torque is related to the motor horsepower by the equation

$$\text{torque in pound-feet} = \frac{\text{hp} \times 5252}{\text{r/min}}$$

For example, the torque of a 50-hp motor running at 1725 r/min is computed as follows:

$$\text{torque} = \frac{50 \times 5252}{1725} = \text{approximately } 152 \text{ lb-ft}$$

If 200 lb-ft is required to drive a particular load, the 50-hp motor just described will be overloaded and will draw a current in excess of the *full-load current* (FLC), which is the current required to produce full-load torque at the rated speed.

8-1.4   *Motor Temperature Ratings*

Both *ambient temperature* and *temperature rise* must be considered when planning electric motor systems.

The *ambient temperature* is the temperature of the air where the motor and control equipment are operating. Most motor controllers are of the enclosed type, and the ambient temperature is the temperature of the air *outside the enclosure,* not inside. Similarly, if a motor has an ambient temperature of 30°C (86°F), this is the temperature of the air outside the motor. Except in very special applications, motors and controllers manufactured to NEMA standards are subject to a 40°C (104°F) ambient temperature limit.

No matter what ambient temperature exists when the motor starts, the *temperature rises* after the motor is running. This temperature increase is because of the current passing through the motor windings. The difference between the winding temperature of the motor (when running) and the ambient temperature is called the *temperature difference.*

The temperature rise produced at full load is not harmful, provided that the motor ambient temperature does not exceed 40°C (104°F). Higher temperature caused by increased current, or higher ambient temperatures, produce a deteriorating effect on motor insulation and lubrication.

A rule of thumb states that for each increase of 10° F above the rated temperature, motor life is cut in half.

8-1.5 *Motor Service Factor*

If the motor manufacturer has given a motor a *service factor,* it means that the motor can be allowed to develop more than the rated or nameplate horsepower without causing undue deterioration of the insulation. The service factor is a margin of safety. For example, if a 10-hp motor has a service factor of 1.15, the motor can be allowed to develop 11.5 hp, although there is no assurance that the motor will develop more than 10 hp.

The service factor is a matter of motor design. NEMA standards for motors list service factors of 1.15 to 1.25 for general-purpose a-c motors from $\frac{1}{2}$ to 200 hp. Other motors, such as totally enclosed, fan-cooled, and motors over 200 hp, have a standard 1.0 service factor. Because of the way in which the standards are established, there is no simple way of determining the service factor without looking at the motor nameplate. Service factors are particularly important when selecting the thermal units of overload relays (Sec. 8-3.2). It is important to know when a motor has a service factor of 1.0, because failure to recognize this fact results in oversized thermal unit selection.

8-1.6 *Motor Duty Ratings (Time Ratings)*

Most motors have a *continuous duty rating.* This permits indefinite operation at a rated load.

Intermittent duty rating is based on a fixed operating time (5, 15, 30, 60 min) after which the motor must be allowed to cool (usually for a specific period of time).

Some motors have both continuous and intermittent duty ratings. Such motors operate continuously at a given load, and intermittently at higher loads.

8-1.7 Plugging, Jogging, and Inching

The term *plugging* is used when a motor running in one direction is momentarily reconnected to reverse the direction and is brought to rest very rapidly. *Jogging* (also known as *inching*) describes the repeated starting and stopping (but not reversing) of a motor at frequent intervals for short periods of time. If either plugging or jogging is to occur more frequently than five times per minute, the starter or controller must be *derated*.

For example, a NEMA size 1 starter has a normal duty rating of $7\frac{1}{2}$ hp at 230 V three-phase (or *polyphase,* which is the term used in connection with motors where anything other than single-phase power is applied). With jogging or plugging, the same starter has a maximum rating of 3 hp. From a design standpoint, if the motor must deliver the full $7\frac{1}{2}$ hp with jogging or plugging, a larger starter must be used. In this case, a NEMA size 2 starter is required. The subject of starter and controller ratings from a design standpoint is discussed in Secs. 8–3 through 8–9.

8-1.8 Sequence (Interlocked) Control

Many processes require a number of separate motors that must be started and stopped in a definite sequence, such as a system of conveyers. When starting up, the delivery conveyor must start first with the other conveyors starting in sequence, to avoid pileup of material. When shutting down, the reverse sequence must be followed with time delays between the shutdowns (except for emergency stops) so that no material is left on the conveyors. This is an example of simple *sequence control* (sometimes known as *interlocked control*).

Separate starters can be used, but is common to build a special controller that incorporates starters for each motor, as well as timers, control relays, and so on, to accomplish the timing sequence. From a design standpoint, such controllers are generally "special-order equipment" and are built to designer's specifications (of timing, sequence, etc.) by the manufacturer. In most present-day systems, motor starters are controlled by electronic devices, usually solid-state digital devices under the control of microprocessors. For a thorough discussion of such equipment, your attention is directed to the author's best-selling *Handbook of Microcomputer-Based Instrumentation and Control* (Englewood Cliffs, N.J.: Prentice-Hall, Inc., 1984).

8-1.9 Locked Rotor Current (LRC)

During the acceleration period at the moment a motor is started, the motor draws a high current called the *inrush current* (similar to the inrush current of a tungsten lamp; see Chapter 7). When the motor is connected directly to the line (so that the full line voltage is applied to the motor), the inrush current is called the *locked rotor current* (LRC) or possibly the *stalled rotor current* (SRC). The LRC can be from about 4 to 10 times the motor full-load current

(FLC; see Sec. 8–3.1). Most motors have an LRC of about six times FLC. The six-times value is often expressed as 600% of FLC.

8–1.10 Motor Controllers, Starters, Switches, Contactors, and Relays

All of the equipment listed in this heading can be considered as *motor control equipment*. The terms are often used interchangeably. Equally often, an improper term is used. For example, contactors and magnetic controllers are often confused. For that reason, the following definitions are established for terms used in this chapter. In turn, the devices listed here are covered in separate paragraphs.

A *motor switch* provides only an on–off function for the motor. However, some motor switches also provide the reversing direction of the motor. Motor switches are discussed further in Sec. 8–6.

A *motor controller* includes some or all of the following functions: starting, stopping, overload protection, overcurrent protection, reversing, changing speed, jogging, plugging, sequence control, and possibly pilot-light indication. A motor controller can also provide control of auxillary equipment, such as breaks, clutches, solenoids, heaters, and signals. A motor controller may be used to control a single motor or a group of motors.

The terms *starter* and *controller* mean practically the same thing. Strictly speaking, a starter is the simplest form of controller, and is capable of starting and stopping the motor, as well as providing the motor with overload protection. The starter shown in Fig. 8–1 can qualify as a controller since it provides on–off (by means of a switch) and overload protection (by means of a *thermal overload element* to the left of the switch).

Motor controllers and starters can be either manual or magnetic. However, the adjective "manual" applies only to starters, whereas both starters and controllers can be magnetic. Manual starters are discussed in Sec. 8–4. Magnetic starters and controllers are covered in Sec. 8–5.

The general classification *contactor* covers a type of magnetically operated device designed to handle relatively *high currents*. As discussed in Chapter 7, a special form of contactor is used for lighting equipment. The lighting contactors discussed in Chapter 7 can handle inrush currents up to about 20 times normal or continuous current, so such contactors are suitable for motor control. However, lighting contactors do not normally provide overload protection. Separate overload protection must be provided if lighting contactors are used to control motor loads. This can also be true if a device is listed by the manufacturer as a contactor or motor contactor. A true starter or controller must include overload protection.

The confusion is further compounded by the fact that a motor contactor is identical in appearance, construction, and current-carrying ability to the equivalent NEMA-size magnetic starter or controller. The magnet assembly and coil, contacts, holding circuit interlock, and other structural features are

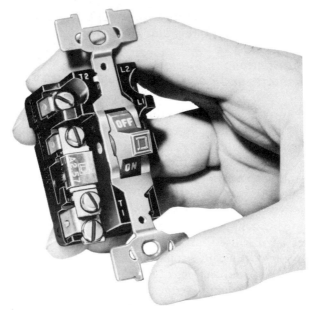

FIGURE 8-1 Starter switch includes overcurrent protection. (Courtesy Square D Company.)

the same, except that true contactors do not include overload protection, as must starters and controllers.

Contactors generally have from one to five poles, as do starters and controllers. Although normally open (NO) and normally closed (NC) contacts can be provided, the great majority of applications use the NO contact configuration, and there is little (if any) conversion of contact operation in the field. Contactors are discussed further in Sec. 8-7.

A *control relay* is also a magnetically operated device, similar in operating characteristics (but not appearance) to a contactor. However, the relay is used to switch low-current circuits. Typically, relays are used in *control circuits,* where little current is needed (on the order of 15 A at 600 V maximum), whereas contactors are used in the *power circuit* (line voltage directly to motor windings), where heavy current is needed.

In a typical application, the coil of a relay is operated by a switch (say an ON–OFF switch), with the relay contacts being used to control the coil of a contactor (or possibly a starter/controller). In turn, the contacts of the contactor are used to control the heavy current between line and the motor load. It is also possible for the coil of a contactor (or starter/controller) to be operated directly by a switch.

Some relays used in motor control have as many as 10 to 12 poles, with various combinations of NO and NC contacts. Also, some relays have convertible contacts, permitting changes to be made in the field from NO to NC operation, or vice versa, without requiring kits or additional components. Control relays are discussed further in Sec. 8-8.

Control relays are not to be confused with *overload relays* discussed in Sec. 8–3. Overload relays are motor overload control devices, usually part of or used with motor starters and controllers.

8-1.11 *Motor Control Equipment Enclosures*

The NEMA and other organizations have established standards for enclosure construction of motor control equipment. Generally, these are the same standards applied to circuit breaker enclosures discussed in Sec. 3–9, and are provided for the same reasons (to prevent accidental contact with live parts, protect the control from harmful environmental conditions, and prevent explosion or fires that might result from the electrical arc caused by the control). Common types of enclosures for motor control equipment are:

NEMA 1: General Purpose (see Fig. 8–2). Usually, the NEMA 1 is used where the exact type of enclosure is not specified. If specific conditions are given, the designer must use the corresponding type of enclosure. The rest of this list summarizes common NEMA types. The NEMA standard literature must be consulted for a full description of the types. However, the titles for each type are generally self-explanatory.

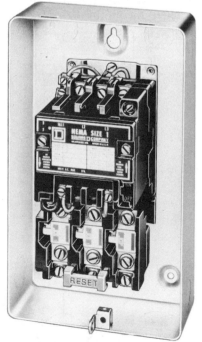

FIGURE 8-2 NEMA 1 general-purpose enclosure. (Courtesy Square D Company.)

NEMA 3: Dust-tight, raintight.

NEMA 3R: Rainproof, sleet resistant.

NEMA 4: Watertight.

NEMA 4X: Watertight, corrosion resistant.

NEMA 7: Hazardous locations, Class I (meets NEC Class I hazardous location standards).

NEMA 9: Hazardous locations, Class II (meets NEC Class II hazardous location standards).

NEMA 12: Industrial use.

NEMA 13: Oiltight, dust-tight.

8-2 WIRING DIAGRAMS USED IN MOTOR CONTROL CIRCUITS

Before going into the details of motor control characteristics, let us discuss the basics of wiring diagrams used in motor control circuits. Both *wiring diagrams* and *elementary* or *schematic diagrams* are used.

Wiring diagrams, such as shown in Fig. 8–3, illustrate (as closely as possible) the actual physical location of all component parts in the system. In the case of the circuit in Fig. 8–3, the system shown is for a motor starter. The dashed lines represent power circuit connections made to the starter, and from the starter to the motor.

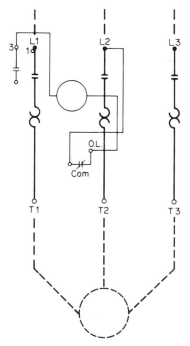

FIGURE 8-3 Typical wiring diagram. (Courtesy Square D Company.)

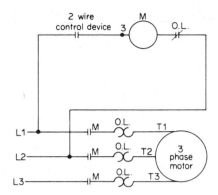

FIGURE 8-4 Typical elementary (schematic) diagram. (Courtesy Square D Company.)

Since wiring connections and terminal markings are shown, the wiring diagram is helpful when wiring the starter, or when tracing wires when troubleshooting. Note that bold lines denote the power circuit, and thin lines are used to show the control circuit. Conventionally, in a-c magnetic equipment, black wires are used in power circuits, and red wiring is used for control circuits.

A wiring diagram is limited in its ability to convey a clear picture of the sequence of operation of a controller, starter, and so on. Where an illustration of the circuit in its simplest form is desired, the elementary diagram or schematic diagram is used.

The *elementary* or *schematic diagram* (sometimes known as a *line diagram*), such as shown in Fig. 8-4, gives a fast, easily understood picture of the circuit. The devices and components are not shown in their actual positions. Instead, the arrangement of components is designed to show the sequence of operation of the devices, and helps in understanding how the circuit operates. The effect of operating various control devices can be readily seen. This helps in troubleshooting, particularly with the more complex controllers.

A summary of the symbols used in wiring diagrams and schematic diagrams is given in Fig. 8-5.

SUPPLEMENTARY CONTACT SYMBOLS

SPST N.O.		SPST N.C.		SPDT		TERMS
SINGLE BREAK	DOUBLE BREAK	SINGLE BREAK	DOUBLE BREAK	SINGLE BREAK	DOUBLE BREAK	SPST – SINGLE POLE SINGLE THROW
						SPDT – SINGLE POLE DOUBLE THROW
						DPST – DOUBLE POLE SINGLE THROW
DPST. 2 N.O.		DPST. 2 N.C.		DPDT		
SINGLE BREAK	DOUBLE BREAK	SINGLE BREAK	DOUBLE BREAK	SINGLE BREAK	DOUBLE BREAK	DPDT – DOUBLE POLE DOUBLE THROW
						N.O. – NORMALLY OPEN
						N.C. – NORMALLY CLOSED

FIGURE 8-5 Standard elementary diagram symbols. (Courtesy Square D Company.)

The diagram symbols shown below have been adopted by the Square D Company and conform where applicable to standards established by the National Electrical Manufacturers Association (NEMA).

SWITCHES

DISCONNECT	CIRCUIT INTERRUPTER	CIRCUIT BREAKER W/THERMAL O.L.	CIRCUIT BREAKER W/MAGNETIC O.L.	CIRCUIT BREAKER W/THERMAL AND MAGNETIC O.L.	LIMIT SWITCHES		FOOT SWITCHES	
					NORMALLY OPEN	NORMALLY CLOSED	N.O	N.C
					HELD CLOSED	HELD OPEN		

PRESSURE & VACUUM SWITCHES		LIQUID LEVEL SWITCH		TEMPERATURE ACTUATED SWITCH		FLOW SWITCH (AIR, WATER, ETC.)	
N.O.	N.C.	N.O.	N.C.	N.O.	N.C.	N.O.	N.C.

SPEED (PLUGGING)	ANTI-PLUG	SELECTOR

SPEED (PLUGGING): F, R — ANTI-PLUG: F, R

SELECTOR

2 POSITION

	J	K
A1	1	
A2		1

1-CONTACT CLOSED

3 POSITION

	J	K	L
A1	1		
A2		1	

1-CONTACT CLOSED

2 POS. SEL. PUSH BUTTON

	SELECTOR POSITION			
CONTACTS	A		B	
	BUTTON		BUTTON	
	FREE	DEPRES'D	FREE	DEPRES'D
1-2	1			1
3-4		1	1	

1-CONTACT CLOSED

FIGURE 8-5 (Cont.)

PUSH BUTTONS									PILOT LIGHTS		
MOMENTARY CONTACT						MAINTAINED CONTACT		ILLUMINATED	INDICATE COLOR BY LETTER		
SINGLE CIRCUIT		DOUBLE CIRCUIT	MUSHROOM HEAD	WOBBLE STICK		TWO SINGLE CKT.	ONE DOUBLE CKT.		NON PUSH-TO-TEST	PUSH-TO-TEST	
N.O.	N.C.	N.O. & N.C.									

CONTACTS								COILS		OVERLOAD RELAYS		INDUCTORS
INSTANT OPERATING				TIMED CONTACTS—CONTACT ACTION RETARDED AFTER COIL IS:				SHUNT	SERIES	THERMAL	MAGNETIC	IRON CORE
WITH BLOWOUT		WITHOUT BLOWOUT		ENERGIZED		DE-ENERGIZED						
N.O.	N.C.	N.O.	N.C.	N.O.T.C.	N.C.T.O.	N.O.T.O	N.C.T.C.					AIR CORE

TRANSFORMERS					AC MOTORS				DC MOTORS			
AUTO	IRON CORE	AIR CORE	CURRENT	DUAL VOLTAGE	SINGLE PHASE	3 PHASE SQUIRREL CAGE	2 PHASE 4 WIRE	WOUND ROTOR	ARMATURE	SHUNT FIELD	SERIES FIELD	COMM. OR COMPENS. FIELD
										(SHOW 4 LOOPS)	(SHOW 3 LOOPS)	(SHOW 2 LOOPS)

FIGURE 8-5 (*Cont.*)

253

8-3   PROTECTING ELECTRIC MOTORS

Motors can be damaged, or their effective life reduced, when subjected to continuous current only slightly higher than the full-load current (FLC; Sec. 8-1.3), times the service factor (Sec. 8-1.5). Motors are designed to handle inrush or locked rotor currents (LRC; see Sec. 8-1.9) without excessive temperature rise, provided that the acceleration time is not too long or the duty cycle (see "jogging" in Sec. 8-1.7) too frequent. Damage to insulation and motor windings can also be sustained on extremely high currents of short duration, as found in grounds or short circuits.

All currents in excess of FLC can be classified as overcurrents. However, the NEC makes a distinction based on the magnitude of the overcurrent and the equipment to be protected.

An overcurrent up to LRC is usually the result of *mechanical overload* on the motor. The subject of protection against this type of overcurrent is covered in *Article 430 (Part C) of the NEC, entitled "Motor Running Overcurrent (Overload) Protection."* In this chapter, the designation is shortened to "overload protection" and is covered in Sec. 8-3.2.

Overcurrents resulting from short circuits or grounds are much higher than the LRC. Equipment used to protect against damage due to this type of overcurrent must not only protect the motor, but also the branch-circuit conductors and the motor controller. Provisions for such protective equipment are specified in *NEC Article 430 under Part D, entitled "Motor Branch Circuit Short-Circuit and Ground-Fault Protection."* In this chapter, the designation is shortened to "overcurrent protection" and is covered in Sec. 8-3.1.

8-3.1   *Overcurrent Protection and Conductor Ampacity*

The function of the overcurrent protective device is to protect the motor branch-circuit conductors, control devices, and motor from shorts or grounds. As in the case of other branch circuits, the protective devices most commonly used for motors are fuses and circuit breakers. One limitation with fuses is that they do not provide a disconnect means, so fuses must be used with switches. Both magnetic-only (instantaneous trip) and thermal-magnetic breakers (Chapter 3) are used. However, the magnetic-only breakers are used *only in combination* with controllers or starters (per NEC Article 430), since magnetic-only breakers have instantaneous (magnetic) tripping action, not the inverse-time tripping (time delay) characteristics of thermal-magnetic breakers. Combination starters and controllers are discussed in Sec. 8-9.

From a design standpoint, the difficulty in selecting the correct size of overcurrent protection for motor branch circuits is the difference between starting and temporary-overload currents, and normal full-load current. If the overcurrent device is selected to protect at the FLC, the device will trip or blow when starting or when temporary overloads occur. If the overcurrent device is selected on the basis of starting or temporary-overload currents, the branch circuit is not protected from prolonged overloads.

If fuses are used to protect the motor branch circuit, this problem can be overcome by means of time-delay (dual-element) fuses such as Class K or R. These fuses have a time delay before they blow (such as a 10-s delay at 500% of rated current). The same results can be achieved with thermal-magnetic circuit breakers since they are not instant-trip, as are magnetic-only breakers.

Full-Load Current Ratings. No matter what device is used, overcurrent protection size or rating is based on the motor FLC. In turn, the FLC is based on motor horsepower (hp) and voltage used. *The NEC includes tables showing the motor FLC for various horsepower ratings and voltage. These values should be used to calculate overcurrent protection, as well as controller or starter size and conductor ampacity.* Figure 8–6 shows the FLC for some typical electrical motors. These FLC values can be used as guidelines, but should also be checked against the NEC of latest issue.

Fuse Selection. Figure 8–7 lists some recommended K5 or RK5 fuse sizes to be used *for three-phase motors only.* These recommendations are based on optimum coordination with Square D overload relays (Sec. 8–3.2) using standard trip thermal units. Figure 8–7 also provides a means of finding the correct motor starter size. Figure 8–7 is quite simple to use. As an example, assume that the motor to be protected from overcurrents has a 50-hp rating and that the service is 230 V, three-phase. Simply go down the 230-V column of Fig. 8–7 until you find a motor hp of 50. The recommended K5 or RK5 fuse size is 200 A, and the recommended starter is NEMA size 4. If the service is increased, say to 575 V, the fuse size is reduced to 80 A, and a NEMA size 3 starter can be used.

Thermal-Magnetic Circuit Breaker Selection. Note that some versions of the NEC refer to thermal-magnetic breakers as "time limit breakers" while other versions use the term "inverse time" breakers. Inverse time describes the function of the thermal-magnetic breaker more accurately, since the trip time is inversely proportional to the amount of current it takes to trip the breaker. Figure 8–8 shows the *maximum* thermal-magnetic circuit breaker rating to be used for *all types of electric motors*. Again, these recommendations are for use with Square D equipment. *If the values listed in Fig. 8–8 exceed that allowed by NEC, the NEC limit must be observed, unless a higher value is necessary to permit motor starting, per NEC 430.* Also, for *part-winding motor starters,* two circuit breakers must be used, one for each motor winding. Maximum rating for each breaker is one-half the value shown in Fig. 8–8. Note that if FLC for any motor is less than 6.67 A, the smallest available circuit breaker (15 A) may be used.

Horsepower ratings of squirrel-cage and synchronous motors with code letters B to E operating at usual speeds							Thermal-magnetic nonadjustable inverse-time circuit breakers	
Three-phase 60 Hz				Single-phase 60 Hz		Full load (A)	Ordinary service	Heavy service
200 V	230 V	460 V	575 V	115 V	230 V			
—	—	—	$\frac{1}{2}$	—	—	0.8	15FA	15FA
—	—	$\frac{1}{2}$	—	—	—	1.0	15FA	15FA
—	—	—	$\frac{3}{4}$	—	—	1.1	15FA	15FA
—	—	$\frac{3}{4}$	1	—	—	1.4	15FA	15FA
—	—	1	—	—	—	1.8	15FA	15FA
—	$\frac{1}{2}$	—	—	—	—	2.0	15FA	15FA
—	—	—	1.5	—	—	2.1	15FA	15FA
$\frac{1}{2}$	—	—	—	—	—	2.3	15FA	15FA
—	—	1.5	—	—	—	2.6	15FA	15FA
—	—	—	2	—	—	2.7	15FA	15FA
—	$\frac{3}{4}$	—	—	—	—	2.8	15FA	15FA
$\frac{3}{4}$	—	—	—	—	—	3.22	15FA	15FA
—	—	2	—	—	—	3.4	15FA	15FA
—	1	—	—	—	—	3.6	15FA	15FA
—	—	—	3	—	—	3.9	15FA	15FA
1	—	—	—	—	—	4.14	15FA	15FA
—	—	3	—	—	—	4.8	15FA	15FA
—	1.5	—	—	—	—	5.2	15FA	15FA
1.5	—	—	—	—	—	5.98	15FA	15FA
—	—	—	5	—	—	6.1	15FA	15FA
—	2	—	—	—	—	6.8	15FA	15FA
—	—	—	—	—	$\frac{3}{4}$	6.9	15FA	15FA
—	—	—	—	$\frac{1}{3}$	—	7.2	15FA	15FA
—	—	5	—	—	—	7.6	15FA	15FA
2	—	—	—	—	—	7.82	15FA	15FA
—	—	—	—	—	1	8.0	15FA	20FA
—	—	—	7.5	—	—	9.0	15FA	20FA
—	3	—	—	—	—	9.6	20FA	20FA
—	—	—	—	$\frac{1}{2}$	—	9.8	20FA	20FA
—	—	—	—	—	1.5	10.0	20FA	20FA
3	—	7.5	10	—	—	11.0	20FA	25FA
—	—	—	—	—	2	12.0	25FA	25FA
—	—	—	—	$\frac{3}{4}$	—	13.8	25FA	30FA
—	—	10	—	—	—	14.0	25FA	30FA
—	5	—	—	—	—	15.2	30FA	35FA
—	—	—	—	1	—	16.0	30FA	35FA
—	—	—	15	—	3	17.0	35FA	35FA
5	—	—	—	—	—	17.5	35FA	35FA
—	—	—	—	1.5	—	20.0	40FA	40FA
—	—	15	—	—	—	21.0	40FA	45FA
—	7.5	—	20	—	—	22.0	45FA	45FA
—	—	—	—	2	—	24.0	50FA	50FA
7.5	—	—	—	—	—	25.3	50FA	50FA
—	—	20	25	—	—	27.0	60FA	60FA
—	10	—	—	—	5	28.0	60FA	60FA
10	—	—	30	—	—	32.0	60FA	70FA
—	—	25	—	3	—	34.0	70FA	70FA
—	—	30	—	—	7.5	40.0	80FA	80FA

FIGURE 8-6 FLC for some typical electrical motors.

| Horsepower ratings of squirrel-cage and synchronous motors with code letters B to E operating at usual speeds | | | | | | | Thermal-magnetic nonadjustable inverse-time circuit breakers | |
| Three-phase 60 Hz | | | | Single-phase 60 Hz | | | | |
200 V	230 V	460 V	575 V	115 V	230 V	Full load (A)	Ordinary service	Heavy service
—	—	—	40	—	—	41.0	80FA	90FA
—	15	—	—	—	—	42.0	80FA	90FA
15	—	—	—	—	—	48.3	90FA	100FA
—	—	—	—	—	10	50.0	90FA	100FA
—	—	40	50	—	—	52.0	90FA	110KA
—	20	—	—	—	—	54.0	90FA	110KA
—	—	—	—	5	—	56.0	90FA	110KA
20	—	—	60	—	—	62.0	100FA	125KA
—	—	50	—	—	—	65.0	100FA	150KA
—	25	—	—	—	—	68.0	100FA	150KA
—	—	60	75	—	—	77.0	110KA	175KA
25	—	—	—	—	—	78.2	110KA	175KA
—	30	—	—	7.5	—	80.0	110KA	175KA
30	—	—	—	—	—	92.0	125KA	200LA
—	—	75	—	—	—	96.0	125KA	200LA
—	—	—	100	—	—	99.0	150KA	200LA
—	—	—	—	10	—	100.0	150KA	200LA
—	40	—	—	—	—	104.0	150KA	225LA
40	—	—	—	—	—	120.0	175KA	250LA
—	—	100	—	—	—	124.0	200KA	250LA
—	—	—	125	—	—	125.0	200KA	250LA
—	50	—	—	—	—	130.0	200KA	300LA
—	—	—	150	—	—	144.0	200LA	300LA
50	—	—	—	—	—	150.0	200LA	300LA
—	60	—	—	—	—	154.0	225LA	350LA
—	—	125	—	—	—	156.0	225LA	350LA
60	—	—	—	—	—	177.0	250LA	400MA
—	—	150	—	—	—	180.0	250LA	400MA
—	75	—	200	—	—	192.0	250LA	400MA
75	—	—	—	—	—	221.0	300LA	450MA
—	—	200	—	—	—	240.0	350LA	500MA
—	—	—	250	—	—	242.0	350LA	500MA
—	100	—	—	—	—	248.0	350LA	500MA
100	—	—	—	—	—	285.0	400LA	600MA
—	—	—	300	—	—	289.0	400LA	600MA
—	—	250	—	—	—	302.0	400LA	700MA
—	125	—	—	—	—	312.0	450MA	700MA
—	—	—	350	—	—	336.0	500MA	700MA
125	—	—	—	—	—	359.0	500MA	800MA
—	150	—	—	—	—	360.0	500MA	800MA
—	—	300	—	—	—	382.0	500MA	800MA
150	—	350	—	—	—	414.0	600MA	900NH
—	—	—	500	—	—	472.0	700MA	1000NH
—	—	400	—	—	—	477.0	700MA	1000NH
—	200	—	—	—	—	480.0	700MA	1000NH
200	—	—	—	—	—	552.0	800MA	1200NH
—	—	500	—	—	—	590.0	800MA	1200NH
—	250	—	—	—	—	602.0	800MA	1200NH

FIGURE 8-6 (*Cont.*)

Minimum starter size (NEMA)	200 V		230 V		460 V		575 V	
	Motor HP	Fuse A	Motor HP	Fuse A	Motor HP	Fuse A	Motor HP	Fuse A
00					$\frac{1}{2}$	1.6	$\frac{1}{2}$	1.25
	$\frac{1}{2}$	3.5	$\frac{1}{2}$	3.2	$\frac{3}{4}$	2.25	$\frac{3}{4}$	1.8
	$\frac{3}{4}$	5	$\frac{3}{4}$	4.5	1	2.8	1	2.25
	1	6.25	1	5.6	1.5	4	1.5	3.2
	1.5	10	1.5	8	2	5.6	2	4.5
0	2	12	2	10	3	8	3	6.25
	3	17.5	3	15	5	12	5	10
1	5	25	5	25	7.5	17.5	7.5	15
	7.5	40	7.5	35	10	20	10	17.5
2					15	35	15	25
			10	40	20	40	20	35
	10	50	15	60	25	50	25	40
3	15	70	20	80	30	60	30	50
	20	90	25	100	40	80	40	60
	25	110	30	125	50	100	50	80
4					60	110	60	90
	30	125	40	150	75	150	75	110
	40	175	50	200	100	175	100	150
5	50	225	60	225	125	225	125	175
	60	250	75	300	150	250	150	200
	75	300	100	350	200	350	200	300
6					250	450	250	350
	100	400	125	450	300	500	300	400
	125	500	150	500	350	600	350	450
	150	600	200	600	400	600	400	500

FIGURE 8-7   Recommended K5 or RK5 fuse sizes to be used for three-phase motors only.

NEMA size	Trip rating (A)
00	15
0	225% of motor FLC
1	225% of motor FLC
1P	225% of motor FLC
2	225% of motor FLC
3	200% of motor FLC
4	175% of motor FLC
5	175% of motor FLC
6	175% of motor FLC

FIGURE 8-8 Maximum thermal-magnetic circuit breaker rating to be used for all types of electric motors.

The values shown in Fig. 8-8 are *maximum*. Figure 8-6 shows *recommended thermal-magnetic trip ratings*. These recommended ratings are for average conditions and are based on trip characteristics of Square D circuit breakers and the NEC tables for inverse-time (thermal-magnetic) breaker requirements of squirrel cage and synchronous motors with code letters B to E, inclusive, with a trip rating not to exceed 200% of the FLC. Lower trip ratings may be required for motors with code letter A.

Higher trip ratings may be required for motors without code letters or with code letters F to V, inclusive. Under some conditions, the next-size-larger breaker trip may be necessary to accommodate the starting of the motor, and is permitted according to NEC. However, in no case may the maximum ratings in Fig. 8-8 be exceeded.

Also note that there are two recommended trip rating columns in Fig. 8-6, one for ordinary service and one for heavy service. *Ordinary service* is for normal starting duty only, with acceleration time of 10 s or less. Always use Square D recommendations for longer acceleration times. *Heavy service* is for jogging or plugging duty, or cycling load with over 25 starts per hour, or over five starts per minute.

Figure 8-6 is equally simple to use. For example, assume that our 50-hp motor is to be protected by a thermal-magnetic circuit breaker in the same 230-V three-phase system. Look up 50 hp in the 230-V three-phase column of Fig. 8-6, and find that the FLC is 130 A, that a 200KA breaker (200 A) is recommended for ordinary service, and that a 300LA breaker (300 A) is recommended for heavy service. If the service is increased to 575 V, the recommended breaker size is reduced to 90 FA (90 A) for ordinary service, and to 110KA (110 A) for heavy service. The FLC is also reduced to 52 A with a 575-V service.

Magnetic-Only Circuit Breaker Selection. Magnetic-only circuit breakers, referred to as "instantaneous trip" circuit breakers by the NEC, have instantaneous (magnetic) tripping action. *Article 430 of the NEC states: "An instantaneous trip circuit breaker shall be used only if adjustable, if part of a* *combination controller having motor-running overload, and also short- circuit and ground-fault protection in each conductor, and if the combination is especially approved for the purpose."* Square D provides MAG-GARD (adjustable, instantaneous-trip, magnetic-only; see Sec. 3–4.4) circuit breakers in their combination controllers. This provides branch-circuit short-circuit protection which is recognized by UL and complies with NEC requirements. Combination starters and controllers are discussed in Sec. 8–9.

MAG-GARD circuit breakers have a single adjustment which simultaneously sets the magnetic trip levels of each individual pole. *The NEC lists the maximum setting of instantaneous-trip breakers as 700% of a-c motor FLC. However, Article 430 of NEC permits this setting to be increased up to a maximum of 1300% if necessary to allow starting of the motor.* The smallest MAG-GARD rating is 3 A, with an adjustable trip range of 7 to 22 A, so close protection can be provided for fractional-horsepower motors. Optional thermal-magnetic circuit breakers are available in Size 0, 1, and 2 combination controllers (Secs. 8–5 and 8–9).

Figure 8–9 shows the *maximum* instantaneous-trip circuit breaker to be used with most electric motors. Again, these recommendations are for use with all Square D combination starters, except part-winding, autotransformer, and wye-delta starters. The NEC 1300% maximum setting may be inadequate for instantaneous-trip circuit breakers to withstand current surges typical of the magnetization current of *autotransformer-type reduced voltage starters,* or *open-transition wye-delta starters* during transfer from "start" to "run," and of constant-horsepower multispeed motors. Use the thermal-magnetic circuit breakers listed in Figs. 8–6 and 8–8 for such applications. *Part-winding motors,* per NEC 430, should have two circuit breakers selected from Fig. 8–9, at not more than one-half the allowable trip setting for the horsepower rating. The two breakers should operate simultaneously as a disconnecting means per NEC 430.

To use Fig. 8–9, first find the FLC from Fig. 8–6. Then select a maximum value in the "motor full-load amps" column of Fig. 8–9. As an example, assume that our 50-hp motor is to be protected by a magnetic-only (instantaneous-trip) breaker (used as part of a combination starter or controller) in the same 230-V three-phase system. Figure 8–6 shows that the FLC of a 50-hp motor is 130 A using 230-V three-phase. Figure 8–9 shows that either a KAL 36225-26M or KAL 36225-29M Square D breaker can be used. The KAL 36225-29M is probably the most practical choice since it allows for an FLC of 133 A. Also note that the -29M breaker has a rating of 225 A (less than 200% of the FLC) and has an adjustable trip rating of 875 to 1750 A, permitting the trip to be set at the NEC recommended 700% of FLC, on up to slightly over the 1300% NEC recommendation.

Conductor Ampacity.   As in the case of other branch circuits, the conductor ampacity in a motor circuit must be equal to (or greater than) the protective device. For example, if a 200-A circuit breaker is used to protect a

NEMA starter size	Thermal unit type	Rating (A)	Adjustable trip range (A)	MAG–GARD catalog number	Motor full load (A)	
					Min.	Max.
0	B or AR only	3	7–22	FAL 36003–11M	0.54	1.91
	B, FB, or AR	7	18–58	FAL 36007–12M	1.65	5.04
		15	50–150	FAL 36015–13M	4.34	12.0
1	B or AR only	3	7–22	FAL 36003–11M	0.54	1.91
		7	18–58	FAL 36007–12M	1.65	5.04
		15	50–150	FAL 36015–13M	4.34	13.0
		30	100–300	FAL 36030–15M	11.3	26.0
		50	150–480	FAL 36050–16M	22.5	26.0
2	B, FB, or AR	15	50–150	FAL 36015–13M	3.85	13.0
		30	100–300	FAL 36030–15M	11.3	26.0
		50	150–480	FAL 36050–16M	22.5	41.7
		100	330–1200	FAL 36100–18M	35.9	45.0
3	CC, FB, or AU	50	150–480	FAL 36050–16M	22.5	41.7
		100	330–1200	FAL 36100–18M	35.9	86.0
4	CC, FB, or AU	100	330–1200	FAL 36100–18M	21.2	86.9
		225	625–1250	KAL 36225–25M	74.8	108
		225	750–1500	KAL 36225–26M	93.5	130
		225	875–1750	KAL 36225–29M	113	133
5	DD or AF	225	875–1750	KAL 36225–29M	67.3	152
		225	1000–2000	KAL 36225–30M	131	173
		225	1125–2250	KAL 36225–31M	150	195
		400	1250–2500	LAL 36400–32M	169	217
		400	1500–3000	LAL 36400–33M	187	260
		400	1750–3500	LAL 36400–35M	225	266
6	B or AR	400	1500–3000	LAL 36400–33M	116	267
		400	1750–3500	LAL 36400–35M	225	304
		400	2000–4000	LAL 36400–36M	262	347
		600	2500–5000	MAL 36600–40M	300	434
		600	3000–6000	MAL 36600–42M	374	520
		600	3500–7000	MAL 36600–44M	449	520

FIGURE 8-9 Maximum instantaneous-trip circuit breaker to be used with most electric motors.

motor that draws 130 A FLC, the conductors must be capable of carrying the 200 A.

8-3.2 Overload Protection

A motor has no intelligence and will attempt to drive any load, even if excessive. Except for inrush or LRC when accelerating, the current drawn by the motor when running is proportional to the load, varying from *no load* current (approximately 40% of FLC) to the FLC rating stamped on the motor nameplate. When the load exceeds the torque rating of a motor, the motor draws higher than FLC. This condition is described as an overload. The maximum overload exists under LRC, in which the load is so excessive that the motor stalls or fails to start, and draws continual inrush current or LRC.

Overloads can be electrical as well as mechanical in origin. Single phasing (Sec. 3-2.6) of a polyphase motor, and low line voltage, are examples of electrical overload.

Overload Protection Basics.   The effect of an overload is a rise in temperature in the motor windings. The larger the overload, the more quickly the temperature increases, to a point damaging to the insulation and lubrication of the motor. An inverse relationship exists between current and time; the higher the current, the shorter the time before motor damage or burnout can occur.

All overloads shorten motor life by deteriorating the insulation. Relatively small overloads of short duration cause little damage, but can be just as harmful as heavy overloads, if sustained. The relationship between the magnitude (percent of full load) and duration (time in minutes) of an overload is shown in Fig. 8-10.

The ideal overload protection for a motor is an element with current-sensing properties (very similar to the heating curve of a motor) which act to open the motor circuit when FLC is exceeded. Operation of the protective device should be such that the motor is allowed to carry harmless overloads, but is quickly removed from the line when an overload has persisted too long.

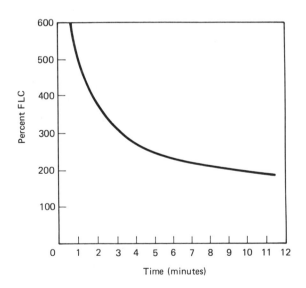

Application of motor heating curve data: On 300% overload, the particular motor for which this curve is characteristic reaches the permissible temperature limits in 3 minutes. Overheating or motor damage occurs if the overload persists beyond that time.

FIGURE 8-10 Typical motor heating curve.

Overload Protection with Fuses and Circuit Breakers. It is possible (but not very practical) that dual-element or time-delay fuses could provide motor overload protection. However, fuses have the disadvantage of being nonrenewable and must be replaced. This problem could be overcome by using thermal-magnetic breakers for motor overload. However, the branch circuit must be designed so that each motor is protected by a separate breaker, and the breaker cannot be used to protect any other load (lighting, resistance, etc.) which is again not very practical. These problems can be overcome by means of overload relays.

Overload Protection with Overload Relays. Overload relays provide individual protection for motors. There are two basic types of overload relays, *thermal* and *magnetic*. Only the thermal type is discussed here, since magnetic overload relays react only to current excesses and are not affected by temperature. The thermal overload relay has inverse trip time characteristics, permitting the relay to hold in during the accelerating period (when inrush current is drawn), yet providing protection on a small overload above FLC when the motor is running. Like a circuit breaker, the overload relay is renewable and can withstand repeated tripping and reset cycles without need of replacement.

Overload relays are intended to protect motors, controllers, and branch-circuit conductors against excessive heating due to prolonged motor overcurrents, up to and including locked rotor currents. It must be emphasized that overload relays do not provide short-circuit protection, which is the function of overcurrent (Sec. 8–3.1) protective equipment such as fuses and circuit breakers.

The overload relay consists of a current-sensing unit connected in the line to the motor, plus a mechanism (actuated by the sensing unit) which serves to directly or indirectly break the motor circuit. In a manual starter (Sec. 8–4), an overload trips a mechanical latch, causing the starter contacts to open and disconnect the motor from the line. In magnetic starters (Sec. 8–5), an overload opens a set of contacts within the overload relay itself. These contacts are wired in series with the starter coil in the control unit of the magnetic starter. Breaking the coil circuit causes the starter contacts to open, disconnecting the motor from the line.

Thermal Overload Relay Basics. Thermal overload relays sense motor current by converting this current to heat in a resistance element. The heat generated is used to open a normally closed contact in series with a starter coil, causing the motor to be disconnected from the line. In spite of being relatively simple and inexpensive, thermal overload relays are very effective in providing motor-running overcurrent protection. This is because the most vulnerable part of a typical motor is the winding insulation, and this insulation is very susceptible to damage by excessively high temperature. Being a thermal model of a motor, the thermal overload relay produces a shorter trip time at a higher current, similar to the way a motor reaches the temperature limit in a shorter time at a higher current. Similarly, in a high ambient temperature, a thermal overload relay trips at a lower current, or vice versa, allowing the motor to be used up to its maximum capacity in the ambient temperature (provided that the motor and overload relay are at the same ambient temperature). Once tripped, the thermal overload relay does not reset until the relay has cooled, automatically allowing the motor to cool before restart.

Square D manufactures two basic types of thermal overload relays, the *melting alloy* and the *bimetallic*. In some types, the bimetallic is available in both *noncompensated* and *ambient temperature-compensated* versions. In both melting alloy and bimetallic, single-element and three-element overloads are available.

With the exception of Types CO, TO, and UO, all Square D thermal overloads incorporate a *trip-free reset mechanism* which allows the relay to trip on an overload, even though the reset lever is blocked or held in the reset position. This mechanism also prevents the control circuit contact from being reclosed until the overload relay and the motor have cooled.

Hand-Reset Melting-Alloy Overload Relays. These relays, shown in Fig. 8–11, use a eutectic alloy solder which responds to the heat produced in a heater element by the motor current. When tripped, the overload relay may be reset manually, after allowing a few minutes for the motor and relay to cool and the solder to solidify. The one-piece thermal unit construction of Square D overload relays provide overload protection for the majority of motors. Each thermal unit is factory tested to ensure trip accuracy. Repeated tripping does not affect the original calibration. Melting alloy thermal units are available in three designs: *standard trip, slow trip,* and *quick trip.*

Standard trip (Class 20) thermal units provide trip characteristics for normal motor acceleration up to approximately 7 s on a full-voltage start.

Slow trip (Class 30) thermal units (Type SB) provide trip characteristics for motor acceleration up to approximately 12 s on a full-voltage start. (The motor should be suitable for extended starting periods.)

Quick trip (Class 10) thermal units (Type FB) are used to protect hermetically sealed, submersible pump, and other motors that can endure locked rotor current for a very short time, or motors that have a low ratio of locked rotor to full-load current.

The contact modules of Type S, F, and G (Series C only) overload relays are replaceable.

In those cases where it is necessary to signal an overload condition, the overload relay is fitted with a set of contacts that close when the relay trips, thus completing an alarm circuit. These contacts are called *alarm contacts.* Contact modules with one normally open and one normally closed contact are available for Type S, F, and G (Series C only) overload relays.

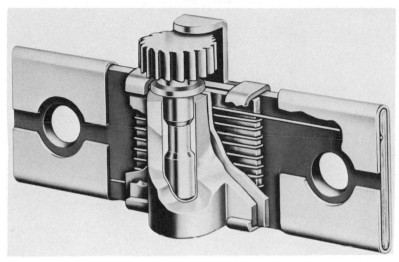

(a)

FIGURE 8-11 Hand-reset melting-alloy overload relays: (a) standard trip melting alloy thermal unit; (b) Type SEO-5 three-pole construction. (Courtesy Square D Company.)

(b)

Automatic-Reset Bimetallic Overload Relays. These relays, shown in Fig. 8-12, are used where the controller is remote, unattended, or difficult to reach. Three-wire control (Sec. 8-5.6) is normally required for safety when used on automatic reset. Two-wire control (Sec. 8-5.5) is not normally used for automatic reset. If two-wire control is used with automatic reset, when the overload relay contacts reclose after tripping, the motor restarts. Unless the cause of the overload is removed, the overload relay trips again. The cycle repeats and, eventually, the motor burns out as a result of the accumulated

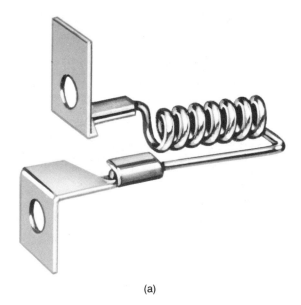

(a)

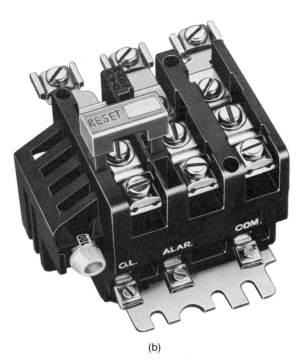

(b)

FIGURE 8-12   Automatic-reset bimetallic overload relays:
(a) bimetallic thermal unit; (b) Type SEO-6B2 three-pole construction. (Courtesy Square D Company.)

heat from the repeated inrush current. More important is the possibility of danger to personnel. The unexpected restarting of a machine when the operator or maintenance person is attempting to find out why the machine has stopped places the person in a hazardous situation.

Normally, bimetallic relays are used on applications that require automatic reset. However, bimetallics can be easily converted in the field to hand-reset, and vice versa. Square D bimetallic overload relays are shipped from the factory set for hand-reset operation. When used on hand-reset, allow the motor and thermal units a few minutes to cool before resetting.

The trip current of bimetallic units is adjustable from 85 to 115% of the trip current rating. A single-pole double-throw contact is standard only on Type S relays. The normally open contact can be used in an alarm circuit.

Ambient-Temperature-Compensated Bimetallic Overload Relays. Where the motor is in a nearly constant ambient temperature, and the controller is in a varying ambient, Square D recommends that ambient-temperature-compensated bimetallic overload relays be used. Such compensation is available on Type S relays only. With Type S, an additional bimetallic unit maintains a nearly constant trip current in overload relay ambient temperatures from −20 to +165°F. In addition to ambient temperature compensation, these overload relays have all the features of the automatic reset bimetallic.

Selecting Thermal Units for Overload Relays. From a design standpoint, motor FLC, the type of motor, and the possible difference in ambient temperature must all be taken into account when choosing overload relay thermal units or overload heaters. Motors of the same horsepower and speed do not all have the same FLC. The NEC includes tables that list FLC according to motor horsepower and voltage. These tables must be used, rather than the motor nameplate FLC, to determine the ampacity of conductors, ampere ratings of switches, or branch-circuit overcurrent devices (as discussed in Sec. 8-3.1). However, *Article 430 of the NEC specifically states that "separate motor-running overcurrent (overload) protection shall be based on the motor nameplate current rating."* Square D furnishes an instruction sheet with every starter.

Each instruction sheet includes thermal unit selections, and either fuse or circuit breaker selections, or both as applicable. Square D also publishes booklets describing overload and short-circuit (overcurrent) selection for their products. These recommendations must be used for Square D equipment, since they allow the designer to achieve proper coordination with a minimum effort by taking advantage of Square D engineering experience. No matter what equipment or manufacturer is used, proper coordination requires a thorough knowledge of the time versus current limits of all branch circuit components, as well as the time versus trip characteristics of the overload relay and short-circuit protective device. To sum up, follow the manufacturer's recommendations. (When all else fails, follow instructions.)

Relationship of Controller and Overload Relay Ratings.   NEMA standards require that controllers be able to interrupt currents up to 10 times FLC. Therefore, it is acceptable for the overload relay to respond before the short-circuit protector up to FLC. At currents above 10 times FLC, the short-circuit protective device must respond first to minimize equipment damage. A fully coordinated system is achieved when the overload relays operate in response to motor overloads before the fuses or circuit breakers, and the fuses or circuit breakers open the circuit before the overload relays trip or burn out on short-circuit currents.

From a design standpoint, another major concern of any overload relay is that the coil operates at the available voltage and the contact carry the current. Usually, the contacts are rated for both alternating current and direct current. For example, a relay may have a continuous current rating of 10 A, a make rating of 7200 VA (volt-amperes), and a break rating of 720 VA. If the voltage through the contacts is 240 V, the instantaneous current at the time contact is made is about 30 A (7200 VA/240 V), whereas the instantaneous break current is about 3 A (720 VA/240 V). With the motor operating continuously, the contacts can carry 10 A. Figure 8-13 shows the contact ratings for Square D thermal overload relays.

NEC versus Manufacturer's Recommendations.   Minimum safety provisions for the control of motors are set forth in the NEC. Although these *minimum provisions must be met,* they are no substitutes for an intelligent selection of protective devices made on the basis of the motor circuit being designed. *The NEC recognizes this fact in Article 430, which states: "Where maximum branch-circuit protective device ratings are shown in the manufacturer's overload relay table for use with a motor controller or are otherwise marked on the equipment, they shall not be exceeded even if higher values are allowed (by the code)."*

Relationship of Motor Heating to Overload Trip Time.   Both melting-alloy and bimetallic overload relays are designed to approximate the heat actually generated in the motor. As the motor temperature increases, so does the temperature of the thermal unit in the relay. The motor and relay heating curves, such as those of Fig. 8-14, show this relationship, in that no matter how high the current drawn, the overload relay provides protection (yet the relay does not trip unnecessarily).

Alternating-current ratings: 50 or 60 Hz

| Type | Volts | Inductive 35% power factor | | | | | Resistive 75% power factor: make, break, and continuous (A) |
| | | Make | | Break | | Continuous (A) | |
		A	VA	A	VA		
AF, AG, AO, AR, AT, AU	120	60	7200	6	720	10	10
C, F, G, T, U, SAF, SAG, SAU	240	30	7200	3	720	10	10
	480	15	7200	1.5	720	10	10
SD0-18 (NEMA A600 a-c ratings)	600	12	7200	1.2	720	10	10
SD and SE (sizes 0 4 and 6–8)	120	30	3600	3	360	6	6
Class 9998	240	15	3600	1.5	360	6	6
Types S0–1 and S0–2	480	7.5	3600	0.75	360	6	6
Contact modules (NEMA B600 a-c ratings)	600	6.0	3600	0.6	360	6	6

Direct-current ratings

| Type | Form | Volts | Inductive and resistive | |
			Make and break (A)	Continuous (A)
AF, AG, AR, AT, AU, SAF, SAG	Std.	120	1.1	10
SAU		240	0.55	10
C, G0–1,	Std.	120	1.1	10
T, U	Y34	240	0.55	10
F, G0–11	Std.	120	0.55	10
SD0–18		240	0.28	10
	Y34	120	0.28	10
		240	0.14	10
AO, SD, and SE (sizes 0–4 and 6–8)	Std.	120	—	—
	Y34	240	—	—

FIGURE 8-13 Contact ratings for Square D thermal overload relays.

8-4 MANUAL STARTERS FOR ELECTRIC MOTORS

Basically, a manual starter is an ON–OFF switch with built-in overload relay thermal units. The small fractional-horsepower unit shown in Fig. 8-1 is a manual starter. With such a unit, the contact mechanism is operated by a mechanical linkage from a toggle handle, which, in turn, is operated by hand. Some manual starters use a pushbutton rather than a toggle switch. With either type, the contact mechanism is also operated by the overload relay or thermal

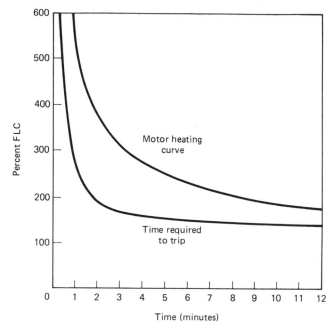

FIGURE 8-14 Motor heating curve versus trip characteristics of overload.

unit. Moving the handle to ON, or pushing the START button, closes the contacts, which remain closed until the handle is moved to OFF, or the STOP button is pressed, or the overload relay thermal unit trips the link mechanism.

Manual starters are generally used on small machine tools, fans and blowers, pumps, compressors, and conveyers, since they are the lowest in cost of all motor starters. Manual starters also have a simple mechanism, and provide quiet operation with no "magnet hum." However, manual starters *cannot* provide low-voltage protection or low-voltage release as can magnetic starters or controllers (Sec. 8–5). If the power fails, the contacts of a manual starter *remain closed,* and the motor restarts when the power returns. This is an advantage for pumps, fans, compressors, oil burners, and the like, but for other applications the automatic restart can be a disadvantage, or even dangerous to personnel or equipment.

From a design standpoint, a manual starter should not be used in an application of any type where the machine or operator could possibly be endangered if power fails and then returns without warning. For such applications, a magnetic starter or controller and momentary-contact pilot device such as described in Sec. 8–5 should be used for safety purposes.

Manual starters are of the *fractional-horsepower* type or the integral-horsepower type. Both types provide *across-the-line* starting. That is, the switches and overload element are in series with the power line (or conductors) to the motor.

8-4.1   Fractional-Horsepower Manual Starter

Fractional-horsepower (FHP) manual starters are designed to control (and provide overload protection for) motors of 1 hp or less, on 120- or 240-V single-phase power. FHP starters are available in single-pole and two-pole versions, and are operated by a toggle handle on the front. Typical wiring diagrams are shown in Fig. 8–15.

When a serious overload occurs, the thermal unit trips to open the starter contacts, disconnecting the motor from the line. In this sense, FHP manual starters are similar to circuit breakers. (However, starters cannot be used as a substitute for circuit breakers or fuses, since starters do not provide branch-circuit overcurrent protection from shorts or grounds. Only combination starters discussed in Sec. 8-9 can provide such protection.)

The contacts of a manual starter cannot be reclosed until the overload device is reset by moving the handle to the "full off" position, after allowing about 2 min for the thermal unit to cool. The open type of manual starter (Fig. 8–1) fits into a standard outlet box and can be used with a standard flush plate.

8-4.2   Integral-Horsepower Manual Starter

The integral-horsepower manual starter is available in two-pole and three-pole versions, to control single-phase motors up to 5 hp and polyphase motors up to 10 hp. A typical integral-horsepower manual starter is shown in Fig. 8–16.

Two-pole starters have one overload relay; three-pole starters usually have two overload relays, but are available with three overload relays. When an overload relay trips, the starter mechanism unlatches, opening the contacts

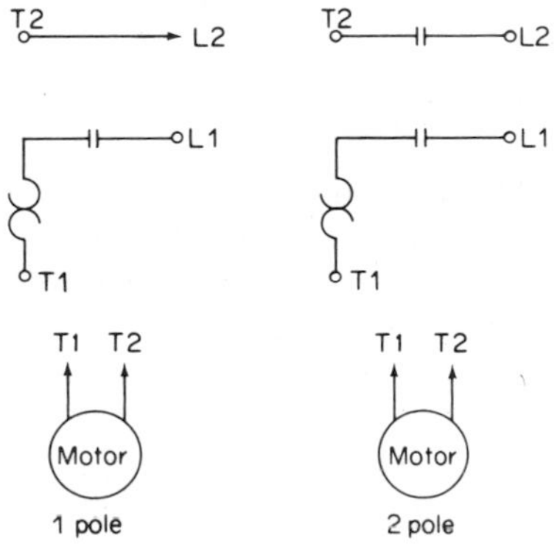

FIGURE 8-15 Typical manual starter wiring diagram. (Courtesy Square D Company.)

FIGURE 8-16 Typical integral horsepower manual starter. (Courtesy Square D Company.)

to stop the motor. The contacts cannot be reclosed until the starter mechanism has been reset by pressing the STOP button, or moving the handle to the RESET, after allowing time for the thermal unit to cool.

8-4.3 Manual Motor Starting Switches

Manual motor starting switches provide ON–OFF control of single-phase or three-phase motors where overload protection is *not* required, or is provided separately. Typically, motor starter switches are used where the branch circuits are adequately protected for overcurrents, and the motors are protected individually for overloads. Toggle operation of the manual switch is similar to the FHP starter (Sec. 8-4.1), and typical applications include small machine tools, pumps, fans, conveyors, and other electrical machinery that has separate motor protection.

Manual motor starting switches are often used with resistance heating units and other nonmotor loads that require the corresponding voltage and current ratings for the contacts.

8-5 MAGNETIC CONTROL
FOR ELECTRIC MOTORS

A high percentage of commercial and industrial applications require the controller to be capable of operation from remote locations, or to provide automatic operation in response to signals from such pilot devices as thermostats, pressure or float switches, and limit switches. Magnetic control is required in such applications. A typical magnetic starter (or controller), NEMA size 1, is shown in Fig. 8–17.

With manual control, the starter must be mounted so that the starter is easily accessible to the operator. With magnetic control, the pushbutton stations or other pilot devices can be mounted anywhere on the machine, and

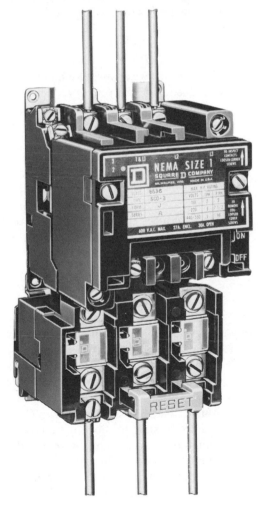

FIGURE 8–17 Typical magnetic starter. (Courtesy Square D Company.)

connected by control wiring into the coil circuit of the remotely controlled starter. However, *NEC requires that a disconnecting means be placed within sight of all motor control locations as discussed in Sec. 3–5.*

8-5.1   Magnetic Starter/Controller Construction

Since we are concerned primarily with design, we shall not discuss construction of magnetic control devices in detail. Instead, we concentrate on those characteristics that most affect design. However, it is necessary to understand the basis of magnetic controller construction to understand the terms.

A typical magnetic controller consists of a magnet assembly, a coil, an armature, and contacts. The armature is controlled by current through the coil. The contacts are mechanically connected to the armature so that when the armature moves to the close position, the contacts close. This applies power to the motor. When the coil is energized and the armature and contacts move to the closed position, the controller is said to be *picked up* and the armature is *seated* or *sealed in.*

The coil draws a fairly high current, known as the *inrush current,* when first energized. Coil current is reduced when the armature and contacts move closer to closing. When the contacts finally close, the current through the coil is known as the *sealed current.* Typically, the inrush current is 6 to 10 times the sealed current.

When the coil is first energized, the impedance is low. After seal-in, the impedance is high. For this reason, a-c magnetic coils should *never be connected in series.* If one contactor seals in ahead of the other (which can occur even if they are identical, and will occur if they are not), the increased circuit impedance reduces the coil current so that the "slow" device does not pick up or, having picked up, does not seal. A-c coils should be connected in *parallel.*

8-5.2   Magnetic Starter/Controller Ratings

Inrush and Sealed Current Ratings.   Magnetic coil information is usually given in units of voltamperes (volts times amperes, VA). For example, for a magnetic starter rated at 600 V inrush and 60 VA sealed, the inrush current of a 120-V coil is 600/120, or 5 A. The same starter with a 480-V coil draws only 600/480, or 1.25 A, inrush, and 60/480, or 0.125 A, sealed.

Magnet Coil Voltage Ratings.   The minimum control voltage that causes the armature to start to move is called the *pickup voltage.* The *seal-in voltage* is the minimum control voltage required to seal the armature. Contact life is extended, and contact damage under abnormal voltage conditions is reduced, if the pickup voltage is also sufficient for seal-in. If the voltage is reduced sufficiently, the controller opens. The voltage at which this happens is called the *drop-out voltage.* Drop-out voltage is somewhat less than seal-in voltage.

Magnet Coil Voltage Variations. Note that the voltage applied to the magnet coil is called the *control voltage* and is separate from the *power voltage* applied through the contacts to the motor.

NEMA standards require that the magnetic device operate properly at varying control voltages, from a high of 110% to a low of 85% of rated coil voltage. This range, established by coil design, ensures that the coil can withstand given temperature rises at voltages up to 10% over rated voltages, and that the armature picks up and seals in, even though the voltage drops to 15% under the nominal rating.

If the voltage applied to the coil is too high, the coil draws more than the design current. Excessive heat is produced and causes early failure of the coil insulation. The magnetic pull is too high, causing the armature to slam home with excessive force. The magnet faces wear rapidly, leading to a shortened life for the controller. Also, contact "bounce" may be excessive, resulting in reduced contact life.

Low control voltages produce low coil currents and reduced magnetic pull. This can result in poor contact closure, chattering of the contacts, arcing, and possible damage to the contacts and coil.

Note that any magnetic device, including magnetic starter coils, produce a characteristic hum. The hum or noise is due mainly to the changing magnetic pull, inducing mechanical vibration. When there is excessive hum, this indicates that the contactor, starter, or relay is defective.

8-5.3 Power Circuits for Magnetic Control Devices

The power circuits of a magnetic control device carry the power from the distribution system to the motor. The power circuits and internal construction of a magnetic starter are shown in Fig. 8-18. As shown, the power circuits of the starter include the stationary and movable contacts and the thermal unit or heater portion of the overload relay assembly. The number of contacts (or "poles") is determined by the electrical service. For example, a three-pole starter is required for a three-phase three-wire system.

From a design standpoint, there are two major concerns in the power circuits of motors. First, the overcurrent devices must sized on the basis of motor current, as discussed in Sec. 8-3.1. In turn, the conductors must be sized on the basis of overcurrent devices, using the information of Chapter 3. Second, the power circuit contacts of the starter or controller must be sized on the basis of motor current. This problem is simplified by means of tables that show electrical ratings for starters and controllers conforming to NEMA standard sizes. A portion of such a table is shown in Fig. 8-19.

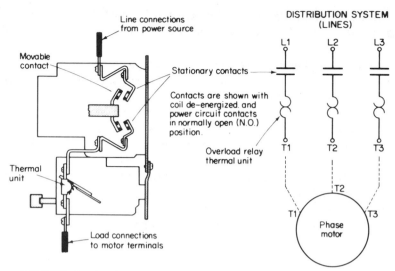

FIGURE 8-18   Power circuits and internal construction of magnetic starter. (Courtesy Square D Company.)

The table of Fig. 8-19 is based on equipment manufactured by Square D Company, but the ratings are directly related to NEMA size. These NEMA sizes take into consideration that the starter contacts must carry the FLC without exceeding a rated temperature rise, and that the contacts are capable of interrupting the motor circuit under LRC conditions.

Basic Starter/Controller Selection Procedure.   Once the motor has been selected, the overcurrent devices sized, and the power wiring sized, the next step is to select the correct starter/controller size. For a given motor application, the starter/controller must equal or exceed the motor horsepower *and* FLC ratings.

For example, assume that we are to select a starter/controller for our old friend the 50-hp motor, using 230-V three-phase power, discussed in Sec. 8-3. Figure 8-6 shows that such a motor has an FLC of 130 A. Confirm this by checking the appropriate NEC table (of latest issue).

By reference to the table of Fig. 8-19, it is seen that a NEMA size 4 starter is required for normal motor duty. If the motor is used for jogging or plugging duty, a NEMA size 5 start is needed.

For a NEMA size 4 at 230 V, the maximum three-phase horsepower rating is 50 (normal duty) and 30 (plugging and jogging duty). The continuous current rating is 135 A, which is 5 A higher than the required 130 A (taken from Fig. 8-6).

For a NEMA size 5 at 230 V, the maximum three-phase horsepower rating is 100 (nornal duty) and 75 (plugging and jogging duty). The continuous current rating is 270 A, well beyond the required 130 A.

NEMA size	Volts	Maximum horsepower rating nonplugging and nonjogging duty		Maximum horsepower rating plugging and jogging duty		Continuous current rating (A) 600 V max.
		Single phase	Poly phase	Single phase	Poly phase	
00	115	$\frac{1}{3}$	—	—	—	9
	200	—	1.5	—	—	9
	230	1	1.5	—	—	9
	380	—	1.5	—	—	9
	460	—	2	—	—	9
	575	—	2	—	—	9
0	115	1	—	$\frac{1}{2}$	—	18
	200	—	3	—	1.5	18
	230	2	3	1	1.5	18
	380	—	5	—	1.5	18
	460	—	5	—	2	18
	575	—	5	—	2	18
1	115	2	—	1	—	27
	200	—	7.5	—	3	27
	230	3	7.5	2	3	27
	380	—	10	—	5	27
	460	—	10	—	5	27
	575	—	10	—	5	27
1P	115	3	—	1.5	—	36
	230	5	—	3	—	36
2	115	3	—	2	—	45
	200	—	10	—	7.5	45
	230	7.5	15	5	10	45
	380	—	25	—	15	45
	460	—	25	—	15	45
	575	—	25	—	15	45
3	115	7.5	—	—	—	90
	200	—	25	—	15	90
	230	15	30	—	20	90
	380	—	50	—	30	90
	460	—	50	—	30	90
	575	—	50	—	30	90
4	200	—	40	—	25	135
	230	—	50	—	30	135
	380	—	75	—	50	135
	460	—	100	—	60	135
	575	—	100	—	60	135
5	200	—	75	—	60	270
	230	—	100	—	75	270
	380	—	150	—	125	270
	460	—	200	—	150	270
	575	—	200	—	150	270
6	200	—	150	—	125	540
	230	—	200	—	150	540
	380	—	300	—	250	540
	460	—	400	—	300	540
	575	—	400	—	300	540
7	230	—	300	—	—	810
	460	—	600	—	—	810
	575	—	600	—	—	810
8	230	—	450	—	—	1215
	460	—	900	—	—	1215
	575	—	900	—	—	1215

FIGURE 8-19 Electrical ratings for motor starters and controllers conforming to NEMA standards.

When starters are recommended by the manufacturer to control other devices, such as lighting, resistance heating, transformer switching, capacitor switching, and so on, tables such as Fig. 8–19 are expanded to show the corresponding ratings. We will not duplicate such information here. The manufacturer's tables (and any corresponding NEC tables) should be consulted, since the tables are subject to change at regular intervals. However, keep the following in mind if a motor starter/controller is used for any purpose other than motor control. The lighting contactors described in Chapter 7 have higher ratings than motor controllers, since lighting contactors must withstand tungsten currents of up to 20 times normal, while motor inrush currents are usually no more than six times normal. Motor starters must be *derated* when used for tungsten lighting, and usually have lower resistance-heating load ratings than corresponding lighting contactors.

8–5.4   *Control Circuits for Magnetic Control Devices*

The control circuits of a magnetic control device carry the current from the start switch, pushbutton, pilot device, and so on, to the magnetic coil of the controller. This circuit to the magnet coil, which causes the starter or controller to pick up and drop out, is distinct from the power circuit (Sec. 8–5.3). Although the power circuit can be single-phase or polyphase, the coil circuit is always a single-phase circuit. Elements of a coil circuit include the following:

1. The magnet coil

2. The contacts of the overload relay assembly

3. A momentary or maintained contact pilot device, such as a pushbutton station, pressure, temperature, liquid level, or limit switch

4. In lieu of a pilot device, the contacts of a relay or timer

5. An auxiliary contact on the starter, designated as a *holding circuit interlock,* which is required in certain control schemes

The coil circuit is generally identified as the *control circuit,* and contacts in the control circuit handle the *coil load.*

The wiring diagram shown in Fig. 8–20a covers the control circuit wiring provided at the factory in a typical magnetic starter. Per NEMA standards, the single-phase control circuit is conventionally wired between lines 1 and 2. As shown in Fig. 8–20b, the control circuit is connected to the single-phase circuit at line 2, but there is no control circuit connection to line 1. Instead, Fig. 8–20b shows that the control circuit is completed by the additional wiring of a pilot device between terminal 3, on the auxiliary contact, and terminal 1 (line 1) on the starter.

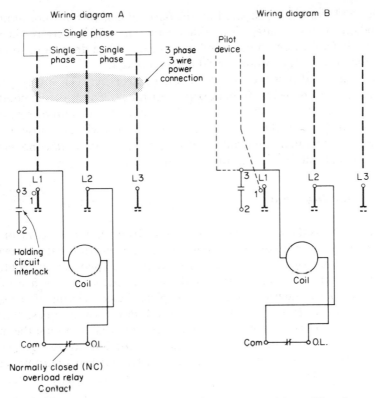

FIGURE 8-20 Typical magnetic starter wiring. (Courtesy Square D Company.)

Control Circuit Currents. Although the power circuit and control circuit voltage may be the same, the current drawn by the motor in the power circuit is much higher than that drawn by the coil in the control circuit. Pilot devices and contacts of timers and relays used in control circuits are not generally horsepower rated, and the current rating is low compared to a starter or contactor.

Inrush and sealed currents of a control circuit can be determined by reference to a magnet coil table (available from manufacturers). As an example, a standard-duty pushbutton with a rating of 15 A inrush, 1.5 A normal (sealed) current at 240 V, 60 Hz, can satisfactorily be used to control the coil circuit of a three-pole NEMA size 3 starter, which has an inrush current of 2.9 A (700 VA/240 V) and a sealed current of 0.2 A (46 VA/240 V).

As a comparison of the differences in current, the power circuit contacts of the same starter may be controlling a 30-hp polyphase motor, drawing an FCL of 78 A.

From a design standpoint, the major concern in control circuit wiring is that the conductors must be sized on the *basis of coil current,* not motor current, and on whatever voltage is used to operate the coil. Generally, the *sealed-in current* is used to determine conductor size, voltage drop, and so on, following the procedures outlined in Chapters 3 and 4.

Keep in mind that there is some voltage drop between the coil and the pilot device. This depends on conductor resistance, length of the conductor, and coil current. If the voltage should drop below the pickup or seal-in voltage, it is possible that the magnetic control contacts will not close when the pilot device is operated. Generally, this is a result of *too-small* conductors for a given coil current and length of conductor run.

8-5.5 Two-Wire Control

Figure 8–21 shows the wiring diagrams for the power circuit and control circuit of a three-phase motor control system. The circuit shown is called a two-wire system, since only two wires (or conductors) are used between the control device (start switch, pushbutton, etc.) and the magnetic starter. When the contacts of the control device close, they complete the coil circuit of the starter, causing the starter to pick up and connect the motor to the lines. When the control device contacts open, the starter is deenergized, stopping the motor.

Two-wire control provides low-voltage release but not low-voltage protection. Wired as shown in Fig. 8–21, the starter functions automatically in response to the direction of the control device, without the attention of an operator.

The dashed portion shown in the elementary diagram represents the *holding circuit interlock* furnished on the starter, but not used in two-wire control. For greater simplicity, this portion is omitted from the conventional two-wire elementary diagram.

8-5.6 Three-Wire Control

Figure 8–22 shows the wiring diagrams for the power circuit and control circuit of a three-phase motor control system, using three-wire control (three wires between control device and magnetic starter).

A three-wire control circuit uses momentary contact start–stop buttons and a holding circuit interlock wired in parallel with the start button to maintain the circuit.

Pressing the normally open start button completes the circuit to the coil. The power circuit contacts in lines 1, 2, and 3 close, completing the circuit to the motor, and the holding circuit contacts (mechanically linked with the power contacts) also close. Once the starter has picked up, the start button can be released, as the now-closed interlock contact provides an alternative current path around the reopened start contact.

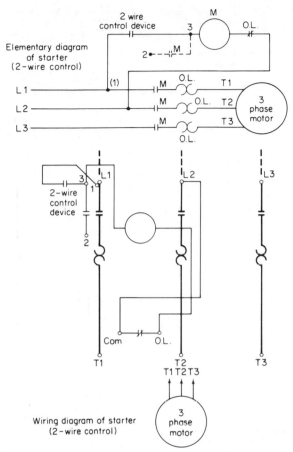

FIGURE 8-21 Two-wire control. (Courtesy Square D Company.)

Pressing the normally closed stop button opens the circuit to the coil, causing the starter to drop out. An overload condition, which causes the overload contact to open, a power failure, or a drop in voltage less than the seal-in value also deenergizes the starter. When the starter drops out, the interlock contact reopens, and both current paths to the coil (through the start button and the interlock) are now open.

8-5.7 Holding Circuit Interlock

The holding circuit interlock is a normally open auxiliary contact provided on standard magnetic starters and controllers. The interlock closes when the coil is energized to form a holding circuit for the starter, after the start button has been released, as described for three-wire control in Sec. 8-5.6.

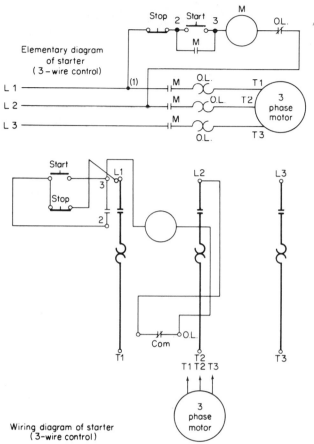

FIGURE 8-22   Three-wire control. (Courtesy Square D Company.)

In addition to the main or power contacts that carry the motor current, and the holding circuit interlock, a starter can be provided with externally attached auxiliary contacts, commonly called *electrical interlocks.* Interlocks are rated to carry *only control circuit currents,* and not motor (power) currents. Normally open and normally closed versions of electrical interlocks are available.

Among a wide variety of applications, interlocks can be used to control other magnetic devices where sequence operation is desired; to electrically prevent another controller from being energized at the same time (such as the reversing starters; see Sec. 8-5.13); and to make-and-break circuits to indicating or alarm devices such as pilot lights, bells, or other signals. Electrical interlocks are packaged in kit form and can easily be added in the field.

8-5.8 Control Device (Pilot Device)

A device operated by some nonelectrical means (such as the movement of a lever) and has contacts in the control circuit of a starter is called a *control device*. Operation of the control device controls the starter and hence the motor. Typical control devices are the switches described in Sec. 6-8.

Some control devices have a horsepower rating and are used to directly control small motors through operation of their contacts. When used in this way, separate overload protection (such as a manual starter) is usually provided, since the control device does not (usually) include overload protection.

A *maintained-contact* control device is one that causes a set of contacts to open (or close) when operated. The contacts stay closed (or open) until a deliberate reverse operation occurs. A conventional thermostat is a typical maintained-contact device. Two-wire control (Sec. 8-5.5) is used with maintained-contact devices.

A standard pushbutton is a typical *momentary-contact* control device. Pushing the button causes normally open contacts to close, and normally closed contacts to open. When the button is released, the contacts revert to their original state. However, the device being controlled (starter, etc.) can be held when a holding circuit interlock (Sec. 8-5.7) and three-wire control (Sec. 8-5.6) are used with a momentary-contact device.

8-5.9 Low-Voltage (Undervoltage) Release

A two-wire control provides *low-voltage release*. The term describes a condition in which a reduction or loss of voltage stops the motor, but in which motor operation automatically resumes as soon as power is restored.

If the two-wire control device in the diagram of Fig. 8-21 is closed, a power failure or drop in voltage below the seal-in value causes the starter to drop out. As soon as power is restored, or the voltage returns to a level high enough to pick up and seal, the starter contacts reclose and the motor runs. This is an advantage in applications involving unattended pumps, refrigeration processes, ventilating fans, and so on.

However, in many applications, the unexpected restarting of a motor after power failure is undesirable. If protection from the effects of a low-voltage condition is required, the three-wire system should be used.

8-5.10 Low-Voltage (Undervoltage) Protection

A three-wire control system such as the one shown in Fig. 8-22 provides *low-voltage protection* (in contrast to low-voltage release of the two-wire system). In either system, the starter drops out and the motor stops in response to a low-voltage condition or power failure.

When power is restored, the starter connected for three-wire control does not pick up, since the reopened holding circuit contact and the normally open start button contact prevent current flow to the coil. To restart the motor after a power failure, the low-voltage protection offered by a three-wire control requires that the start button be pressed. A deliberate action must be performed, ensuring greater safety than that provided by two-wire control.

8–5.11   Full-Voltage (Across-the-Line) Starter

As the name implies, a *full-voltage* or *across-the-line starter* connects the motor to the line directly. All starters and controllers described in this chapter are full-voltage devices.

A motor connected across the line draws full inrush current and develops maximum torque so that the motor accelerates the load to full speed in the shortest possible time. Across-the-line starting can be used whenever this high inrush current and starting torque are not objectionable.

With some loads the high starting torque and high inrush current are not acceptable. In those cases, *reduced voltage starting* is used. The wiring and equipment for such control systems are highly specialized and are not covered in this book.

8–5.12   Common Control versus Transformers and Separate Control

The coil circuit of a magnetic starter or contactor is distinct from the power circuit. The coil circuit can be connected to any single-phase power source, provided that the *coil voltage* and *frequency* match the service to which the coil is connected.

When the control circuit is tied back to lines 1 and 2 of the starter, the voltage of the control circuit is always the same as the power circuit voltage, and the term *common control* is used.

It is sometimes desirable to operate pushbuttons or other control circuit devices at some voltage lower than the motor voltage. In Fig. 8–23, a single-phase control transformer (with dual voltage 240/480-V primary and 120-V secondary) has the 480-V primary connected to the 480-V three-phase three-wire service brought into the starter. The control circuit is supplied by the 120 V secondary. The coil voltage is 120-V, and the pushbuttons or other control devices operate at the same voltage level. A fuse is often used to protect the control circuit (in addition to branch circuit overcurrent protection), and it is common practice to ground one side of the transformer.

Control of a power circuit by a lower control circuit voltage can be obtained by connecting the coil circuit to a *separate control* voltage source rather than to a transformer secondary. Such a separate control system is shown in

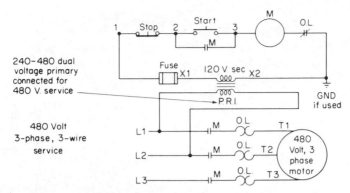

FIGURE 8-23 Wiring for control circuit with transformer. (Courtesy Square D Company.)

Fig. 8-24. The coil rating must match the control source voltage, but the power circuit can be any voltage (typically up to 600 V).

8-5.13 Reversing Starter

Reversing the direction of motor shaft rotation is often required. Three-phase squirrel-cage motors can be reversed by reconnecting *any two* of the three line connectors to the motor. The most practical method for three-phase motor reversal is by means of a reversing starter. A NEMA size 1 three-pole reversing starter is shown in Fig. 8-25. The wiring diagram is shown in Fig. 8-26.

When closed, the forward contacts (F) connect lines 1, 2, and 3 to motor terminals T1, T2, and T3, respectively. As long as the forward contacts are closed, mechanical and electrical interlocks prevent the reverse contacts from being closed.

When the forward-contact coil is deenergized, the reverse-contact coil picks up, closing reverse contacts (R), which reconnect the lines to the motor. Note that by running through the reverse contacts, line 1 is connected to motor terminal T3, and line 3 is connected to motor terminal T1. The motor then runs in the opposite direction.

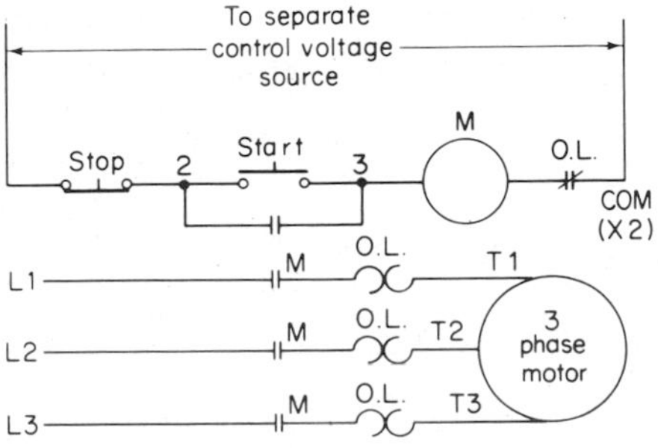

FIGURE 8-24 Wiring for control circuit supplied by separate source (separate control). (Courtesy Square D Company.)

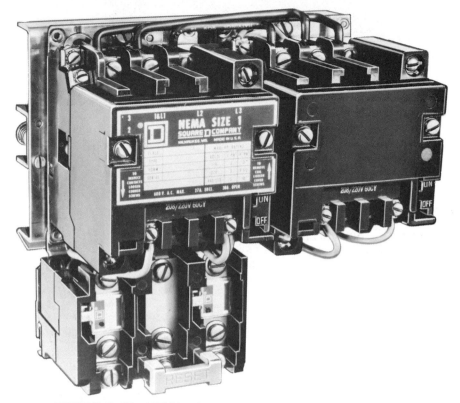

FIGURE 8-25 NEMA size 1 reversing starter. (Courtesy Square D Company.)

The motor-reversing function can also be performed by a contactor. However, no overload protection is provided by a true contactor. Note the three-pole overload relay in the lower left-hand corner of Fig. 8-25. Such a magnetic reversing starter consists of two sets of coils and contacts (suitably interwired), with both electrical and mechanical interlocking to prevent the coils of both units from being energized at the same time.

Manual reversing starters (using two manual starters) are also available. As in the magnetic version, the forward and reverse switching mechanism are

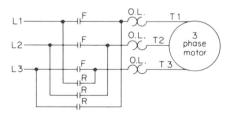

FIGURE 8-26 Wiring diagram for reversing starter. (Courtesy Square D Company.)

mechanically interlocked. However, since coils are not used in the manually operated starters, electrical interlocks are not furnished.

8-6 SWITCHES AS CONTROL DEVICES

Switches are the most common control devices used in motor control circuits. The following is a brief summary of the switch types in most common use.

A *drum switch* is a manually operated three-position three-pole switch used for *manual reversing* of single-phase and three-phase motors. Drum switches are often rated by motor horsepower, in addition to voltage and current.

A *control station* (also known as a pushbutton station) may contain pushbuttons, selector switches, and pilot lights. Figures 8–27 through 8–29 show some typical control stations. Pushbuttons may be momentary-contact or maintained-contact type. Selector switches are usually maintained-contact, or can be of the spring-return type to give momentary-contact operation.

Standard-duty stations (Fig. 8–27) handle the coil currents of starters and controllers up to NEMA size 4. Heavy-duty stations (Figs. 8–28 and 8–29) have higher contact ratings, and provide greater flexibility through a wider variety of controls and interchangeability of units.

A *foot switch* is a control device operated by a foot pedal used where the process or machine requires that the operator have both hands free. Foot switches can be of momentary-contact type or can be mechanically latched.

A *limit switch* is a control device that converts mechanical motion into an electrical control station. Typical applications include start, stop, reverse, slow down, speed up, or recycle machine operations.

Snap switches for motor control purposes are enclosed precision switches that require low operating forces and have a high repeat accuracy.

FIGURE 8-27 Standard duty pushbutton (or control) station. (Courtesy Square D Company.)

FIGURE 8-28   Heavy-duty pushbutton (control) station. (Courtesy Square D Company.)

A *pressure switch* is a control device for pumps, air compressors, welding machines, lubrication systems, and machine tools. Pressure switch contacts are operated by pistons, bellows, or diaphragms against a set of springs. The spring pressure determines the pressure at which the switch closes and opens the contacts.

A *float switch* is a control device where the contacts are controlled by movement of a rod or chain, and a counterweight fitted with a float. Float switches are often used for tank pump motors.

FIGURE 8-29   Heavy-duty oil-tight pushbutton (control) station. (Courtesy Square D Company.)

8-7 MOTOR CONTACTORS

As discussed in Chapter 7, contactors are magnetic control devices used where *high currents* are involved. Motor contactors are essentially the same as the lighting contactors discussed in Chapter 7, but with one major exception. Motor contactors are designed to handle the inrush currents of motors (typically six times FLC, or normal current). Lighting contactors must handle tungsten lighting inrush currents (up to almost 20 times normal current). As a result, lighting contactors of a given size can handle corresponding motor currents, without derating. The reverse is not true. Motor contactors must be derated to handle lighting loads. Also, neither motor nor lighting contactors provide overload protection, which must be provided for motor control (but not for lighting control).

For these reasons, true motor contactors are generally being replaced by motor starter/controllers or by lighting contactors, depending on the application. The following recommendations apply to the great majority of wiring design problems.

If the load is lighting-only, resistance-only, or any combination of lighting, resistance, and motors (or transformers, capacitors, etc.), use a lighting contactor and follow the design recommendations of Chapter 7. Keep in mind that the motor must have separate overcurrent protection if a contactor is used.

If the load is motor-only, or mostly motor, use a motor starter/controller using the design recommendations of this chapter and those of the motor and/or starter manufacturer. Keep in mind that the motor starter/controller has the advantage of not requiring separate overload protection, but must be derated if the load is not motor-only.

8-8 CONTROL RELAYS

A relay is an electromechanical device where the contacts are used in control circuits of magnetic starters, contactors, solenoids, timers, and other relays. Relays are generally used to amplify the contact capability or to multiply the switching function of a pilot device.

Figure 8–30 shows an example of *current amplification*. Relay and starter coil voltages are the same (230 V), but the ampere rating of the temperature switch is too low to handle the current drawn by the starter coil M. A relay is interposed between the temperature switch and the starter coil. The current drawn by the relay coil CR is within the rating of the temperature switch, and the relay contact CR has a rating adequate for the current drawn by the starter coil M.

Figure 8–31 shows an example of *voltage amplification*. A condition may exist in which the voltage rating of the temperature switch is too low to permit

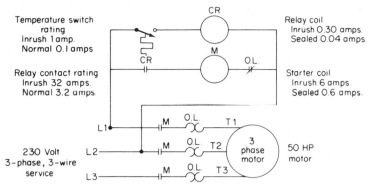

Temperature switch
rating
Inrush 1 amp.
Normal 0.1 amps

Relay contact rating
Inrush 32 amps.
Normal 3.2 amps.

Relay coil
Inrush 0.30 amps.
Sealed 0.04 amps.

Starter coil
Inrush 6 amps.
Sealed 0.6 amps.

230 Volt
3-phase, 3-wire
service

50 HP
motor

3 phase motor

FIGURE 8-30   Relay used for current amplification. (Courtesy Square D Company.)

direct use in a starter control circuit operating at some higher voltage. In this application, relay coil CR and a pilot device (say an on–off switch) are connected to a low-voltage power source (120 V) compatible with the rating of the pilot device and the temperature switch. The relay contact CR, with the higher voltage rating (230 V), is then used to control operation of the starter.

Figure 8-32 shows an example of *multiplying switching functions* of a pilot device with a single or limited number of contacts. As shown, a single-pole pushbutton contact can control operation of many different loads (pilot light, starter, contactor, solenoid, and timing relay), through the use of a six-pole relay. Depressing the ON button energizes the relay coil CR. The normally open contacts close to complete the control circuits to the starter, solenoid, and timing relay. One normally open contact closes to form a holding circuit around the ON button. The normally closed contacts open to deenergize the contactor and turn off the pilot light. Relays are commonly used in complex controllers to provide the logic to set up and initiate proper sequencing and control for a number of interrelated operations.

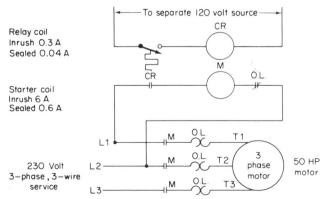

Relay coil
Inrush 0.3 A
Sealed 0.04 A

Starter coil
Inrush 6 A
Sealed 0.6 A

To separate 120 volt source

230 Volt
3-phase, 3-wire
service

50 HP
motor

3 phase motor

FIGURE 8-31   Relay for voltage amplification. (Courtesy Square D Company.)

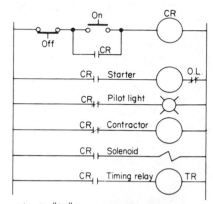

Depressing the "ON" button in this control circuit energies the
relay coil (CR). Its normally open contacts to complete the
control circuits to the starter, solenoid and timing relay, and one
contact forms a holding circuit around the "ON" button. The
normally closed contacts open to de-energize the contactor
and turn off the pilot light.

FIGURE 8-32 Relay used to multiply switching functions.
(Courtesy Square D Company.)

Relays used in motor control differ in voltage ratings (typically 150, 300, and 600 V), number of contacts, contact convertability, physical size, and in attachments to provide accessory functions such as mechanical latching and timing.

In selecting a relay for a particular application, one of the first steps is to determine the *control voltage* at which the relay must operate. Unless there are manufacturer's recommendations to the contrary, use 150-V relays when the service is 115/120 V, 300-V relays when the service is 230/240/277 V, and 600-V relays when the service is 480/575/600 V. Once the voltage is known, the relay having the necessary *contact rating* can be further reviewed, and a selection made on the basis of *number of contacts* and any specialized characteristics needed. Figures 8-33 through 8-35 show typical relays used in motor control, as well as lighting and heating applications.

FIGURE 8-33   150-V relay. (Courtesy Square D Company.)

FIGURE 8-34 300-V relay. (Courtesy Square D Company.)

A *timing relay* is similar to a control relay except that certain contacts of a timing relay are designed to operate as a preset time interval, after the coil is energized or deenergized. When the time interval or delay occurs after the coil is energized, the term *on delay* is used. A delay after the coil is deenergized is called *off delay*.

FIGURE 8-35 600-V relay. (Courtesy Square D Company.)

8-9 COMBINATION STARTERS AND CONTROLLERS

In some applications, combination starter/controllers are used to control electric motors. A combination starter is similar to the combination lighting contactors described in Sec. 7-4.5 in that both devices include disconnect means, overcurrent protection, and starting or contacting functions in one enclosure.

Compared with separately mounted components, the combination starter takes up less space, requires less time to install and wire, and provides greater safety. Personnel safety is assured because the door is mechanically interlocked (the door cannot be opened without first opening the disconnect). Combination starters can be furnished with circuit breakers or fuses to provide overcurrent protection, and are available in nonreversing and reversing versions.

9

BASIC ELECTRICAL DATA
FOR COMMERCIAL AND INDUSTRIAL
WIRING DESIGN CALCULATIONS

This chapter provides a review, or summary, of basic electrical data needed for design of commercial/industrial wiring systems. Basic electrical theory is not included, and complete mathematical analysis of design problems is deliberately avoided. Math is used only where absolutely necessary, and then in the simplest possible form (basic arithmetic, graphic solutions, etc.). The information is included for the student (who needs to understand the problems of design) and for the working electrician or contractor (who needs a ready reference of basic equations, calculation techniques, etc.).

9-1 BASIC OHM'S LAW CALCULATIONS

The calculations for d-c Ohm's law are shown in Fig. 9-1. Calculations for a-c Ohm's law are given in Fig. 9-2. The basic equations for a-c power are essentially the same as for d-c power, except that impedance (Z) is used in place of resistance (R). However, the calculations for a-c and d-c power are different because of the *apparent power* in a-c circuits.

The term "apparent power" is applied when the reactive power factor of an alternating current is disregarded. To obtain true power in an a-c circuit, it is necessary to multiply the apparent power by the cosine of the phase angle. Since the phase angle is always less than 90°, and the corresponding cosine is a fraction of 1, the true power is always less than the apparent power.

The ratio between the true power (generally expressed in watts, or W)

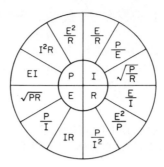

FIGURE 9-1 Direct current Ohm's law equations.

Solution for quantities printed in center section is calculated by substituting known values in the equations adjoining the respective center sections.
For example, if desired quantity is R, and both E and P are known, $R = E^2/P$

and the apparent power (expressed in volt-amperes, or VA) is known as the *power factor*. This can be expressed as

$$\text{power factor} = \frac{E \times I \times \text{cosine phase angle}}{E \times I}$$

or

$$\text{power factor} = \text{cosine phase angle} = \frac{R}{Z}$$

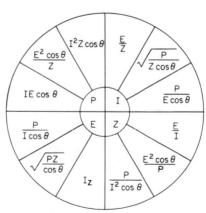

θ = phase angle

FIGURE 9-2 Alternating current Ohm's law equations.

Solution for quantities printed in center section is calculated by substituting known values in the equations adjoining the respective center sections.
For example, if desired quantity is Z, and both E and P are known, $Z = E^2 \cos \theta / P$.

where R is the resistance in an a-c circuit (in ohms) and Z is the impedance (in ohms).

In those a-c circuits where there is no reactance (pure resistance, such as incandescent lighting), the phase angle is zero, so the apparent power is the true power (the power factor is 1).

9-2 RESISTANCE, INDUCTANCE, AND CAPACITANCE IN CIRCUITS

Resistance, either in conductors or in the load, function to control the flow of electrical current and to divide voltages. Although resistances can be used in a variety of circuit combinations, these circuits are versions of the basic series and parallel arrangements shown in Fig. 9-3.

Inductances can be connected in series, parallel, and in series–parallel. Where no interaction of magnetic fields is produced by the inductance, the equations are the same as for resistance, as shown in Fig. 9-4.

When current flows through an inductance, a magnetic field is set up around the inductance. If the current is alternating, the magnetic field varies

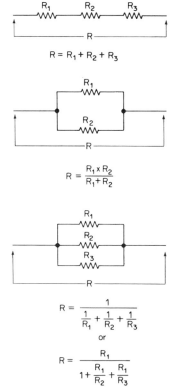

FIGURE 9-3   Resistance in conductors and circuits.

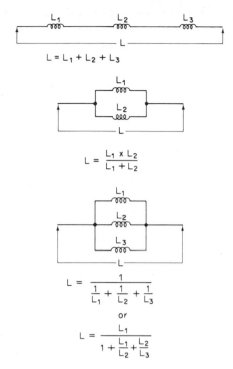

$$L = L_1 + L_2 + L_3$$

$$L = \frac{L_1 \times L_2}{L_1 + L_2}$$

$$L = \frac{1}{\frac{1}{L_1} + \frac{1}{L_2} + \frac{1}{L_3}}$$

or

$$L = \frac{L_1}{1 + \frac{L_1}{L_2} + \frac{L_2}{L_3}}$$

FIGURE 9-4 Inductance in circuits.

(builds up and collapses). The build up and collapse of the magnetic field produces another current in the inductance. This self-induced current is in opposition to the initial or external current. This opposition is known as *inductive reactance* and is expressed in ohms.

The combined effect of the inductance and effective resistance is known as *impedance,* which is also expressed in ohms. The impedance of an inductance can be substituted directly in the basic Ohm's law equations, as shown in Fig. 9–2.

Capacitors function to store an electrostatic charge. Direct current can be stored in a capacitor, but alternating current passes around the capacitor. Capacitors can be used in a variety of circuit combinations. The majority of circuits are a version of the basic series, parallel, or series–parallel arrangements shown in Fig. 9–5.

The *farad* is the basic unit of capacitance, but since the farad is too large for practical applications, submultiples are used. The most common submultiples are the microfarad (μF, one millionth of a farad) and the picofarad (pF, one millionth of one millinth of a farad).

Direct current does not pass through a capacitor, but alternating current does pass around a capacitor. The process of alternately charging and discharging the capacitor produces some opposition. This opposition is called *capacitive reactance* and is expressed in ohms.

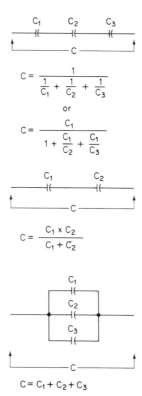

$$C = \cfrac{1}{\cfrac{1}{C_1} + \cfrac{1}{C_2} + \cfrac{1}{C_3}}$$

or

$$C = \cfrac{C_1}{1 + \cfrac{C_1}{C_2} + \cfrac{C_1}{C_3}}$$

$$C = \frac{C_1 \times C_2}{C_1 + C_2}$$

$$C = C_1 + C_2 + C_3$$

FIGURE 9-5 Capacitance in circuits.

The combined effect of the capacitance and resistance is known as impedance, which is also expressed in ohms. The impedance of a capacitor can be substituted directly in the basic Ohm's law equations, as shown in Fig. 9-2.

9-3   BASIC POWER RELATIONSHIPS

A typical electrical system consists of many devices, or loads, of different types and uses. For example, a typical electrical system in an industrial plant can include electric motors, lighting, and resistance heating. It is impractical, if not impossible, to evaluate such loads on the basis of resistance. However, when it is considered that all of these loads are in parallel across a particular voltage, either the total current or the total power can be used to sum up the total power consumption.

The relationships are

$$\text{current} = \frac{\text{power}}{\text{voltage}} \qquad \text{power} = \text{current} \times \text{voltage}$$

For example, if 10 electric motors must be operated with 480 V, and each motor draws a maximum of 10 A, the power is 10 × 10 A = 100 A; 100 A × 480 V = 48,000 VA or 48 kVA. Note that the term VA (voltamperes) is used here instead of W (watts). This is because electric motors usually have some reactance (inductive or capacitive) that must be combined with the resistance to find impedance. The difference between VA and W depends on the power factor. These relationships are best calculated by means of the power triangle, as discussed in Sec. 9–4.

Using another example, if 100 tungsten (incandescent) lamps are operated at 120 V, and each lamp is rated at 60 W, the current is 100 × 60 W = 6000 W or 6 kW; 6kW/120 V = 50 A. Note that watts are used here instead of voltamperes. This is because tungsten or incandescent lamps are considered to have no reactance, but are pure resistance, and the power factor is 1.

9–3.1 Kilowatthour

To be of any practical value, electrical energy must be evaluated as power consumed over a specified period of time. The unit of electrical energy is the watthour. However, since 1 watthour is very small in relation to useful work, the kilowatthour (kWh) can be 1000 W for 1 hour, 1 W for 1000 hours, 10 W for 100 hours, and so on.

Electric utility companies charge for electrical energy on the basis of kilowatthours. The charge depends on such factors as total power consumption, location of facilities, number of meters at the service entrance, demand for a particular period of time, and seasonal conditions.

The relationships are

$$\text{kilowatthour} = \text{power (in kW)} \times \text{time (in hours)}$$

Assume that the 100 electric lamps described in the previous example are to be operated for 8 hours each day, and that the charge is 10 cents per kWh. Find the kWh and the cost per day.

$$\text{kWh} = 6 \text{ kW} \times 8 = 48 \text{ kWh} \qquad \text{cost} = 48 \times \$0.10 = \$4.80$$

9–3.2 Horsepower

Horsepower is the common unit of mechanical power. The watt is the common unit of electrical power. The relationships between watts and horsepower are

$$1 \text{ hp} = 746 \text{ W} \quad \text{or} \quad 0.746 \text{ kW}$$
$$1 \text{ kW} = 1.34 \text{ hp}$$

9-3.3   *Power Efficiency*

Part of the power applied to any circuit or load is consumed by the circuit or load, and is not converted directly into energy. The energy or power output from a load is always less than the power input. The ratio of power output to power input is termed *efficiency* or *power efficiency* and is expressed by

$$\text{efficiency} = \frac{\text{power output}}{\text{power input}}$$

Since output is less than input, the result is a decimal. Multiply the result by 100 to convert into terms of percentage.

For example, assume that an electric motor requires an input of 2.4 kW, or 3.216 hp (2400/746), to produce an output of 3 hp. The efficiency is 0.93, or 93% (3/3.216 = 0.93).

9-4   THE POWER TRIANGLE

A typical electrical distribution system consists of several different loads, all connected in parallel. Some of the loads are pure resistive, such as heating elements and incandescent lamps. Other loads can be part resistive and part inductive, such as electric motors and fluorescent lamps. A typical load is not likely to be pure capacitive, (except for special loads consisting of capacitors (or a synchronous motor) connected to correct the power factor, as discussed in Sec. 9-6.

Unless all the loads are pure resistive, it is not practical to evaluate the total load (or even one of the loads) on the basis of resistance alone. If the total load or any part of the load is inductive, inductive reactance must be included. Since resistance and reactance produce an impedance, the total impedance factor must be considered. And, since impedance is involved, power factor must also be considered.

It is possible to combine all these factors for a single load in an electrical system, or several loads, by means of a *power triangle*, as shown in Fig. 9-6. The power triangle is a graphic means of showing load relationships, and is made up by multiplying the basic impedance triangle by the current through the load. In the example of Fig. 9-6 the load is part resistive and part reactive (inductive) in series. Thus the current is the same in both the reactance and the resistance. When these factors are multiplied by the current, the voltage is found, and the result is a *voltage triangle*. (This is produced by the basic Ohm's law relationship $E = IR$.)

When all the factors in the voltage triangle are again multiplied by current, a power triangle is formed. (This is produced by the basic Ohm's law relationship $P = EI$.) The power triangle has the same shape as the original

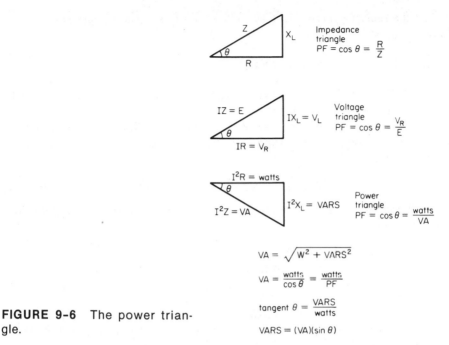

$$VA = \sqrt{W^2 + VARS^2}$$

$$VA = \frac{watts}{\cos\theta} = \frac{watts}{PF}$$

FIGURE 9-6 The power triangle.

$$tangent\ \theta = \frac{VARS}{watts}$$

$$VARS = (VA)(\sin\theta)$$

impedance triangle (same angles and same ratios). However, the horizontal side of the power triangle is expressed in watts (instead of resistance), and the hypotenuse is expressed in voltamperes (VA) (instead of impedance). Another way of expressing this is that the wattage of a power triangle has the same relationship to VA as resistance does to impedance. Or, resistance divided by impedance produces the same fraction or ratio as occurs when wattage is divided by VA. In either case the result is power factor and is equal to the cosine of the phase angle.

Note that the voltamperes unit is sometimes called apparent power since VA is the power that appears to be used ($P = EI$). However, true power is found when VA is multiplied by the power factor. For this reason, it is possible to measure both voltage and current of the load (with a voltmeter and an ammeter) and the wattage of the same load (with a wattmeter) and find that the two do not agree.

The vertical side of the power triangle represents *reactive power* (by itself an imaginary quantity). For this reason, reactive power is expressed in an imaginary electrical unit of *vars* (voltamperes reactive) or *kvars* (1000 voltamperes reactive).

Most power triangles are drawn with the vertical side going down, since most practical electrical loads are inductive, causing the current to lag the voltage. If the load is capacitive, the current leads the voltage, and the vertical side of the triangle is drawn going up. Note that vars is equal to VA × the sine of the phase angle. For this reason, the sine of the phase angle is often called the *reactive factor* (just as the cosine of the phase angle is called the power factor).

9-4.1   Example of Power Triangle

Figure 9–7 shows how the power triangle and the related equations can be used to solve a practical problem.

Suppose that a motor draws 8 A from a 120-V line. The motor is rated at 720 W (just below 1 hp). Find the VA, the power factor, the phase angle, and the vars.

VA: Multiply voltage (120) by current (8) to find a VA value of 960 VA.

Power factor: Divide the wattage (720) by the VA (960) to find a power factor of 0.75.

Phase angle: The cosine of 0.75 is 41.4°, as shown in Fig. 9–8.

Vars: Multiply VA (960) by the sine of 41.4°. Figure 9–8 shows a sine of 0.6613 for 41.4°. So the vars factor is 635 vars (960 × 0.6613 = 635).

A similar problem can be stated in another way. Assume that a motor is rated as 10 hp and operates on 120 V with a power factor of 0.8. What current is drawn?

1. Find the wattage. One horsepower = 746 W; 10 hp = 7460 W.

2. Divide the wattage (7460) by the power factor of 0.8 to find a VA of 9325.

3. Divide the VA (9325) by the voltage (120) to find a current of 77.7 A.

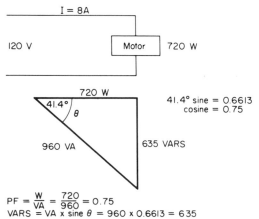

FIGURE 9-7   Example of power triangle calculations.

Angle	Radians	Sine	Cosine	Tangent
0	0.0000	0.0000	1.0000	0.0000
1	0.0175	0.0175	0.9998	0.0175
2	0.0349	0.0349	0.9994	0.0349
3	0.0524	0.0523	0.9986	0.0524
4	0.0698	0.0698	0.9976	0.0699
5	0.0873	0.0872	0.9962	0.0875
6	0.1047	0.1045	0.9945	0.1051
7	0.1222	0.1219	0.9925	0.1228
8	0.1396	0.1392	0.9903	0.1405
9	0.1571	0.1564	0.9877	0.1584
10	0.1745	0.1736	0.9848	0.1763
11	0.1920	0.1908	0.9816	0.1944
12	0.2094	0.2079	0.9781	0.2126
13	0.2269	0.2250	0.9744	0.2309
14	0.2443	0.2419	0.9703	0.2493
15	0.2618	0.2588	0.9659	0.2679
16	0.2793	0.2756	0.9613	0.2867
17	0.2967	0.2924	0.9563	0.3057
18	0.3142	0.3090	0.9511	0.3249
19	0.3316	0.3256	0.9455	0.3443
20	0.3491	0.3420	0.9397	0.3640
21	0.3665	0.3584	0.9336	0.3839
22	0.3840	0.3746	0.9272	0.4040
23	0.4014	0.3907	0.9205	0.4245
24	0.4189	0.4067	0.9135	0.4452
25	0.4363	0.4226	0.9063	0.4663
26	0.4538	0.4384	0.8988	0.4877
27	0.4712	0.4540	0.8910	0.5095
28	0.4887	0.4695	0.8829	0.5317
29	0.5061	0.4848	0.8746	0.5543
30	0.5236	0.5000	0.8660	0.5774
31	0.5411	0.5150	0.8572	0.6009
32	0.5585	0.5299	0.8480	0.6249
33	0.5760	0.5446	0.8387	0.6494
34	0.5934	0.5592	0.8290	0.6745
35	0.6109	0.5736	0.8192	0.7002
36	0.6283	0.5878	0.8090	0.7265
37	0.6458	0.6018	0.7986	0.7536
38	0.6632	0.6157	0.7880	0.7813
39	0.6807	0.6293	0.7771	0.8098
40	0.6981	0.6428	0.7660	0.8391
41	0.7156	0.6561	0.7547	0.8693
42	0.7330	0.6691	0.7431	0.9004
43	0.7505	0.6820	0.7314	0.9325
44	0.7679	0.6947	0.7193	0.9657
45	0.7854	0.7071	0.7071	1.0000
46	0.8029	0.7193	0.6947	1.0355
47	0.8203	0.7314	0.6820	1.0724
48	0.8378	0.7431	0.6691	1.1106
49	0.8552	0.7547	0.6561	1.1504
50	0.8727	0.7660	0.6428	1.1918
51	0.8901	0.7771	0.6293	1.2349
52	0.9076	0.7880	0.6157	1.2799

FIGURE 9-8 Table of trigonometric functions.

Angle	Radians	Sine	Cosine	Tangent
53	0.9250	0.7986	0.6018	1.3270
54	0.9425	0.8090	0.5878	1.3764
55	0.9599	0.8192	0.5736	1.4281
56	0.9774	0.8290	0.5592	1.4826
57	0.9948	0.8387	0.5446	1.5399
58	1.0123	0.8480	0.5299	1.6003
59	1.0297	0.8572	0.5150	1.6643
60	1.0472	0.8660	0.5000	1.7321
61	1.0647	0.8746	0.4848	1.8040
62	1.0821	0.8829	0.4695	1.8807
63	1.0996	0.8910	0.4540	1.9626
64	1.1170	0.8988	0.4384	2.0503
65	1.1345	0.9063	0.4226	2.1445
66	1.1519	0.9135	0.4067	2.2460
67	1.1694	0.9205	0.3907	2.3559
68	1.1868	0.9272	0.3746	2.4751
69	1.2043	0.9336	0.3584	2.6051
70	1.2217	0.9397	0.3420	2.7475
71	1.2392	0.9455	0.3256	2.9042
72	1.2566	0.9511	0.3090	3.0777
73	1.2741	0.9563	0.2924	3.2709
74	1.2915	0.9613	0.2756	3.4874
75	1.3090	0.9659	0.2588	3.7321
76	1.3265	0.9703	0.2419	4.0108
77	1.3439	0.9744	0.2250	4.3315
78	1.3614	0.9781	0.2079	4.7046
79	1.3788	0.9816	0.1908	5.1446
80	1.3963	0.9848	0.1736	5.6713
81	1.4137	0.9877	0.1564	6.3138
82	1.4312	0.9903	0.1392	7.1154
83	1.4486	0.9925	0.1219	8.1443
84	1.4661	0.9945	0.1045	9.5144
85	1.4835	0.9962	0.0872	11.4301
86	1.5010	0.9976	0.0698	14.3007
87	1.5184	0.9986	0.0523	19.0811
88	1.5359	0.9994	0.0349	28.6363
89	1.5533	0.9998	0.0175	57.2900

FIGURE 9-8   *(Cont.)*

9-5   ADDING LOADS WITH THE POWER TRIANGLE

If all the loads in a particular distribution system have the same power factor, it is relatively easy to find the total current of the system. Simply add all the rated wattages, divide by the power factor to find VA, then divide by the system voltage to find the current. However, if the loads have different power factors, it is necessary to use *vectors* to add the wattage. This can be done with the power triangle, as shown in Fig. 9-9.

Assume that there are three 30-kW loads connected in parallel across a 240-V system. The loads have power factors of 0.6, 0.8, and 0.9. Find the total current of the system, as well as the overall power factor for the system.

The 0.6 power factor load has a phase angle (cosine) of approximately

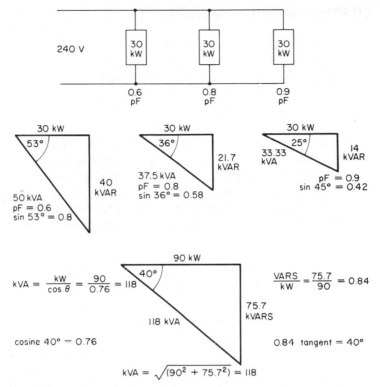

FIGURE 9-9 Adding loads of different power factors with the power triangle.

53° as shown in Fig. 9–8, and a kVA of 50 (30 kW/0.6). The sine of 53° is approximately 0.8. So the load has 40 kvars (50 kVA × 0.8).

The 0.8 power factor load has a phase angle (cosine) of approximately 36° as shown in Fig. 9–8, and a kVA of 37.5 (30 kW/0.8). The sine of 36° is approximately 0.58. So the load has 21.7 kvars (37.5 kVA × 0.58).

The 0.9 power factor load has a phase angle (cosine) of approximately 25° as shown in Fig. 9–8, and a kVA of 33.33 (30 kW/0.9). The sine of 25° is approximately 0.42. So the load has 14 kvars (33.33 × 0.42).

The total kW is 90 (30 + 30 + 30), and the total kvars is 75.7 (40 + 21.7 + 14). The total kvars/total kW ratio is 75.7/90, or approximately 0.84. The tangent 0.84 is for an angle of approximately 40°. The cosine of 40° is approximately 0.76. So the overall power factor is 0.76.

With an overall power factor of 0.76 and a total wattage of 90 kW, the total kVA is approximately 118 (90 kW/0.76). This can be confirmed by the equation

$$VA = \sqrt{W^2 + vars^2}$$

With a kVA of 118 and a system voltage of 240 V, the total current is approximately 493 A.

9-6   CORRECTING THE POWER FACTOR

An increase in power factor for a given electrical system results in lower current, assuming a given voltage and a given wattage. Generally, the voltage and wattage of a power distribution system are specified or are determined by the number of loads involved. Either way, the voltage and wattage cannot be changed. The only way to increase efficiency of a system is to increase the power factor.

Many utility companies require a minimum power factor, typically 0.8 to 0.85 or better. In some cases, the utility companies assess an extra charge if the power factor for a system is below 0.8. This depends on local regulations. Most utility companies are not interested in the power factors of individual loads, but in the overall system power factor, as measured at the service entrance. This is determined when the service entrance wattmeter reading is divided by the VA at the service entrance.

No matter what regulations or conditions exist, it is always good design practice to keep the power factor as high as practical. This can be done in two ways: by adding capacitors across the line or by adding a synchronous motor across the line. Both methods are discussed in this section. No matter which method is used, the basic principle is the same.

Most loads (that are not pure resistive) are inductive and have inductive reactance. If a capacitor is added across the line, the capacitor produces capacitive reactance, even though the capacitor draws no current. The capacitive reactance cancels the inductive reactance, reducing the overall resistance/reactance ratio, and increasing the power factor. A synchronous motor also produces a capacitive effect on the line. Both devices cause the current to lead the voltage. This lead offsets any lag produced by the inductive load, reducing the phase angle to zero (or near zero), and increasing the power factor to 1 (or near 1).

9-6.1   Power-Factor Correction with Capacitors

Assume that a system load is measured at the service entrance. The load is rated at 60 kW. However, the current is 333.33 A at 240 V. This shows an apparent power of 80 kVA (240 V × 333.33 A) and a power factor of 0.75 (60 kW/80 kVA. Find what value of capacitor is required to raise the power factor to 0.85, so the current is reduced to 294 A (60 kW/0.85 = 70.588 kVA; 70.588 kVA/240 V = 294 A).

The angle with a cosine of 0.85 is approximately 32° (Fig. 9–8). The sine of 32° is approximately 0.53. If the desired kVA is 60.588 (as it is if the power factor is 0.85), the required kvars is 0.53 × 70.588, or approximately 37.5 kvars.

The kvars of the existing system (without correction) is equal to the existing VA (80 kVA) times the sine of the existing phase angle. The angle with a cosine of 0.75 (existing) is approximately 41° (Fig. 9–8). The sine of 41° is approximately 0.66. So the existing (lagging kvars) is 0.66 × 80 kVA, or approximately 52.8 kvars.

With an existing 52.8 kvars, and a desired 37.4 kvars, the capacitor must supply 15.4 kvars (leading) (52.8 − 37.4 = 15.4).

Capacitors designed specifically for power-factor correction are rated as to kvars, as well as voltage and frequency. However, it may be necessary to find the capacitance value that produces 15.4 kvars in some cases. The following steps describe the procedure.

Find the leading reactive current (imaginary) that occurs with 15.4 kvars and 240 V; 15.4 kvars/240 V = approximately 64 A (reactive).

With a reactive current of 64 A, and a voltage of 240 V, the capacitive reactance is approximately 3.75 Ω. This is found by

$$\text{capacitive reactance} = \frac{\text{voltage}}{\text{leading reactive current}} = \frac{240}{64} = 3.75$$

The capacitance required to produce a reactance of 3.75 Ω at 60 Hz (the usual line frequency) is approximately 700 μF. This is found by

$$\text{capacitance (in farads)} = \frac{1}{6.28 \times \text{frequency} \times \text{capacitive reactance}}$$

or

$$\frac{1}{6.28 \times 60 \times 3.75} = 0.0007 \text{ F} = 700 \text{ μF}$$

The capacitor must have a voltage rating of at least 240 V, and preferably 1.5 times the line voltage, or 240 V × 1.5 = 360 V.

9-6.2 Power-Factor Correction with Synchronous Motors

When a synchronous motor is connected in parallel with an electrical distribution system, and the field current is properly adjusted in relation to the load, the current drawn by the motor leads the voltage. As far as reactance is concerned, the synchronous motor appears as a capacitor in parallel with the line. So a large synchronous motor driving a fixed load (ventilating fan, etc.) can raise the power factor of the entire system. It is not necessary for the motor to drive a load, although such a design is usually wasteful, since the motor draws some current (even without a load). When a synchronous motor is used for power factor correction, the motor is sometimes called a *synchronous condenser* or *synchronous capacitor*.

Assume that a synchronous motor (without a load) is used to correct the power factor of the system described in Sec. 9–6.1. With such an arrangement, the field current of the motor must be adjusted to draw 64 A (reactive). In theory, the motor produces no power. In practice, the motor produces some power, even if the drive shaft is spinning in free air. This power is added to the overall load, and thus increases the current. However, since the effect of adding leading kvars to the system produces a drastic reduction in current, the load produced by the motor can be neglected in practical design.

Now assume that a synchronous motor (with a 5-kW load) is used to correct the power factor of the system described in Sec. 9–6.1. The original load is 60 kW, 80 kVA, power factor 0.75, 240 V, 333.33 A, phase angle 41°, and 52.8 kvars.

With the motor added, the load is increased to 65 kW. However, the power factor is to be increased to 0.85, with a phase angle (cosine) of 32°. The desired kVA (with a power factor of 0.85) is 76.47 (65 kW/0.85). The sine of 32° is approximately 0.53. If the desired kVA is 76.47, the required kvars is 0.53 × 76.47, or approximately 40.5.

With an existing (before correction) 52.8 kvars and a required (after correction) 40.5 kvars, the synchronous motor must supply 12.3 kvars (leading) (52.8 − 40.5 = 12.3).

The field current of the motor must be adjusted so that the kVA value includes both the 12.3 kvars and the 5-kW load. As shown in Fig. 9–6, $VA = \sqrt{W^2 + vars^2}$. So the motor must be adjusted for a kVA of $\sqrt{5^2 + 12.3^2}$ or approximately 13.3. At 240 V, the current is approximately 55 A (13.3 kVA/240 V = 55 A) for the synchronous motor.

Under these conditions, the overall system current is now 76.47 kVA/240 V, or approximately 315 A. Note that this is about 21 A higher than the 294 A required by the capacitor power factor correction described in Sec. 9–6.1. However, the synchronous motor drives an additional 5-kW load, which requires the additional 21 A.

9-7   THREE-WIRE POWER DISTRIBUTION SYSTEM

The three-wire power distribution system shown in Fig. 9–10 is used in a great majority of commercial electrical systems and in many industrial systems. The three-wire system has the advantage of providing both 120 V and 240 V from the same wiring. The 240 V is used for large loads (such as window air conditioners in offices) to reduce current. All other factors being equal, the current is reduced to one-half when 120 V is used instead of 120 V. The 120 V is used for lighting and small loads.

The utility company generally provides the three wires to the service entrance. However, in very small commercial installations, only two of the wires must be carried throughout the interior wiring. This provides 120 V for the

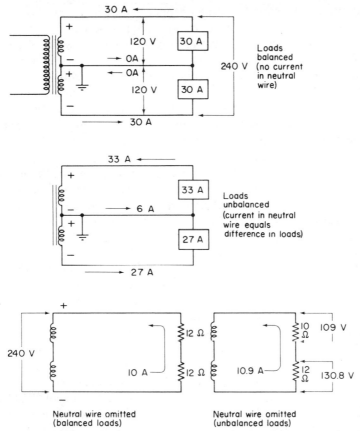

FIGURE 9-10 Three-wire Edison system load arrangements.

interior, and is used only when the customer has no large loads (and none are anticipated). Should the customer add large loads (particularly large air conditioning) the third wire must be carried through the interior from the service entrance. (Local code may require the third wire in some cases.)

When only two wires are used in the interior, one wire must be the *neutral wire*. *This neutral wire must be grounded (at the service entrance; see Chapter 5), must never be fused, and must never be disconnected.*

The system shown in Fig. 9–10 is generally referred to as the Edison three-wire system, and is formed by connecting two 120-V power sources in series so that the polarities cause the voltages to be *additive*. This is shown by the polarity markings of Fig. 9–10. Of course, since ac is involved, the polarity markings are instantaneous. However, at any given instant, the bottom 120-V supply is added to the top 120-V supply so that the total across the two supplies is 240 V. There are several ways of producing such a supply. The most common way is with a center-tapped transformer as shown. The secondary winding is actually two 120-V windings in series.

Ideally, the load should be distributed evenly between the two 120-V supplies. The loads should be the same on either side of the neutral wire. For example, if the total load is 60 A, the branch circuits should be arranged so that 30 A is supplied by each of the 120-V sources. Under these conditions, the system is balanced, and there is no current flowing in the neutral wire. The current through the top load opposes the current through the bottom load in the neutral wire. Since both currents are equal, the currents cancel and produce zero current in the neutral wire.

No matter how well balanced the system is designed, there is no guarantee that all loads are used simultaneously. There is always the possibility of unbalance. When this occurs, current flows in the neutral wire. The amount of current is the difference between the two 120-V load currents. For example, if the load connected to the top 120-V source draws 33 A, and the bottom load draws 27 A (still a total current of 60 A), there is 6 A flowing in the neutral wire (33 A − 27 A = 6 A). From a design standpoint, note that the neutral wire always carries less current than either of the other wires. So the neutral wire size need never be larger than the other wires. In practice, always use the same wire size for all three wires.

Never consider any design where the neutral wire is omitted. Even if all the loads require only 240 V, the neutral wire *must be brought in and grounded at the service entrance.* It is sometimes assumed that if the 120-V loads are balanced, the neutral wire can be omitted, since the neutral does not carry any current, and the voltage drop across each of the loads is equal (a 240-V source divides, with 120 V across each load). This may work in theory, *but not in practice.* If there is the slightest unbalance, the voltages do not divide evenly, resulting in one voltage being too high, and one too low.

As an example, assume that each of the loads are supposed to be 12 Ω, pure resistive, as shown in Fig. 9–10. The total series resistance of the loads is 24 Ω. With 24 Ω and 240 V, the current is 10 A. The 10 A through both loads produces 120 V across each load. Now assume that one load is only 10 Ω, with the other load remaining at 12 Ω. The total load resistance is 22 Ω, and the total current is 10.9 A (240 V/22 Ω). The voltage across the 10-Ω load is 109 V (10 × 10.9); the voltage across the 12-Ω load is 130.8 V (12 × 10.9). The high voltage could cause damage to the equipment. The low voltage can result in poor performance.

9-8   THREE-PHASE POWER DISTRIBUTION SYSTEM

A three-phase generator has the armature coils divided into three sections. The armature coils, regardless of number, are grouped into sets so arranged that the induced voltage for one set of coils differs from the others by a third of a cycle (120°), as shown in Fig. 9–11. Three-phase generators are also known as *polyphase* generators, since they produce more than one phase of alternating current.

Three-phase systems (generators, transformers, motors, etc.) are classified according to connection into two groups: the delta (Δ) connection (so called since the connection resembles the Greek letter) and the wye (Y) or *star* connection. Sometimes the systems are mixed, as when a delta motor is powered by a wye generator through a wye-to-delta transformer.

The main advantage of three-phase systems is that increased power can be obtained without a corresponding increase in individual voltages and currents. So conductor sizes can be kept smaller. In a wye system, this is possible since the line voltage is approximately 1.73 times the voltage across each phase.

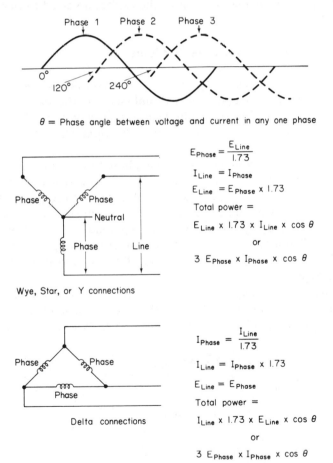

θ = Phase angle between voltage and current in any one phase

$$E_{Phase} = \frac{E_{Line}}{1.73}$$

$$I_{Line} = I_{Phase}$$

$$E_{Line} = E_{Phase} \times 1.73$$

Total power =

$$E_{Line} \times 1.73 \times I_{Line} \times \cos\theta$$

or

$$3\ E_{Phase} \times I_{Phase} \times \cos\theta$$

Wye, Star, or Y connections

$$I_{Phase} = \frac{I_{Line}}{1.73}$$

$$I_{Line} = I_{Phase} \times 1.73$$

$$E_{Line} = E_{Phase}$$

Total power =

$$I_{Line} \times 1.73 \times E_{Line} \times \cos\theta$$

or

$$3\ E_{Phase} \times I_{Phase} \times \cos\theta$$

Delta connections

FIGURE 9-11 Basic connections and calculations for three-phase systems (balanced loads).

In a delta system, the power increase is possible since the line current is approximately 1.73 times the current produced in each phase.

The terms *phase current* (I_{phase}) and *phase voltage* (E_{phase}) refer to currents and voltages in one phase or branch of a three-phase load.

The terms *line current* (I_{line}) and *line voltage* (E_{line}) refer to currents and voltages in one of the lines or connection wires (conductors) between source and load.

Line current is the current found if an ammeter is connected in any one line (conductor). In the case of a wye system, the line current and phase current have the same value.

Line voltage is the voltage found across any two lines (except the neutral conductor in wye). In the case of a delta system, the line voltage and phase voltage have the same value.

The equations of Fig. 9–11 show the factors involved to find the power in both delta and wye systems.

9-8.1   Three-Phase Four-Wire Systems

In certain power systems, where both three-phase and single-phase are required, or where there is a possibility of unbalance in the loads, a neutral wire is used, as shown in Fig. 9–12. The neutral wire is grounded (at the service entrance when the utility company supplies the three-phase four-wire service) and serves two purposes.

First, the neutral wire permits single-phase to be obtained from a three-phase system. For example, there is a 120-V drop across each phase (A, B, and C) shown in Fig. 9–12. This 120 V is obtained with respect to the neutral wire and any one of the other wires. For the system shown, the voltage drop across B is used to supply 120-V single-phase to load circuits R4 and R5. These loads can be lighting, heaters, or other devices requiring only 120-V single-phase. At the same time, a balanced load (R1, R2, and R3) is furnished by the three-phase.

The neutral wire also serves the purpose of carrying the difference current when the loads are not balanced. When all three loads in a three-phase system are in balance (all three loads equal), there is no current in the neutral

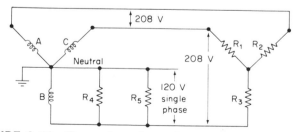

FIGURE 9-12   Three-phase four-wire system, supplying both single-phase (120 V) and three-phase (208 V).

wire. However, if one phase has more (or less) current than the other phases, there is some current in the neutral wire (as discussed in Sec. 9–7).

In the case of the system in Fig. 9–12, the three-phase loads (R1, R2, and R3) are balanced and there is no current in the neutral wire that results from these loads. However, phase B has the additional single-phase loads. So the neutral wire must be capable of carrying the current produced by R4 and R5.

9-8.2 Power in Unbalanced Three-Phase Loads

The equations shown in Fig. 9–12 are based on balanced three-phase loads. When the loads are not balanced, the calculations to find power and current are more difficult and involve the use of *vectors*. The subject of vector calculation is discussed in Sec. 9–9. In this section, we concentrate on finding power and currents in three-phase systems.

Unbalanced Delta.   Figure 9–13 shows the equations and vectors necessary to find the power and line current in unbalanced delta systems. To find total power, the powers in each phase (phase power) must be added. In the

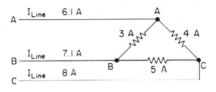

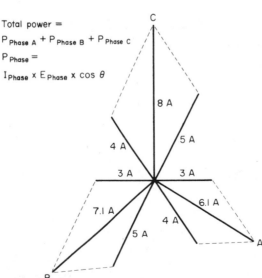

FIGURE 9-13   Power and current in unbalanced delta system.

example shown, the phase currents are 3, 4, and 5 A, and the phase (or line) voltage is 120 V. So the phase powers are 360, 480, and 600 W, and the total power is 1440 W. If the problem is stated in reverse, with the wattage and voltage given, the phase currents are found when the wattage is divided by the voltage (assuming a pure resistive load and a power factor of 1).

Design of a system such as shown in Fig. 9–13 requires that the line current be known, not the phase current. The conductors must carry line current; wire size voltage drop, and so on, must be based on line current.

The current in any one line is the vector sum of the corresponding two-phase currents. For example, the current in line A is the vector sum of the 3-A and 4-A loads; line B is the 3-A and 5-A vector sum; line C is the 4-A and 5-A vector sum. These are relatively simple vectors to find, using the graphic method, when it found that the angle between any of the two-phase currents is 60°.

As shown by Fig. 9–13, the current in line A is 6.1 A, the current in line B is 7.1 A, and the current in line C is 8 A.

Unbalanced Wye.   When there is any possibility of unbalance in a wye system, the neutral wire must be used, so that the voltage across each load is the same (even though the currents may be different). The current in each line is equal to the current in the corresponding phase. If the wattage is given, the current (phase and line) can be found when the wattage is divided by the voltage.

If the wye system is unbalanced, there is some current in the neutral wire. This current is the vector sum of the three lines (or phase) currents. The angle between any two of the currents is 120°, as shown in Fig. 9–14.

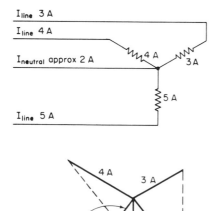

FIGURE 9–14   Current in unbalanced wyte system.

Assume that the values for the system of Fig. 9–14 are the same as for the delta system (3, 4, and 5 A). As shown, the neutral wire current is about 2 A. As a guideline, the current in the neutral wire is less than the smallest line current.

9-9 USING VECTORS AND TRIGONOMETRY

Trigonometry is a branch of mathematics that deals with the properties of triangles, and is useful in a-c electricity because some electrical quantities are best expressed as vectors. A rotating vector (also known as a phasor) is a diagram of lines drawn to scale in the proper direction to indicate the intensity and direction of the forces to be added. Figure 9–15 shows a typical vector, where 1 in. equals 1 V.

Vectors can be solved by two basic methods: (1) the graphic method, where the various lines and angles are laid off on paper using a given scale; and (2) the trigonometric solution. The results of the graphic method are measured physically, using the same scale. The trigonometric method is more exact, but the graphic method is quicker.

Figure 9–15 shows a vector used to combine 10 V and 7.5 V, when the 7.5 V lags the 10 V by 30°. The following is a summary of the graphic solution to such a problem.

Lay off a line 10 in. long, such as *OA* of Fig. 9–15. Then lay off a line *OB* at an angle of 30° to *OA,* but make *OB* approximately 7.5 in. long to indicate 7.5 V. Next, build up a *parallelogram* by making *AC* equal and par-

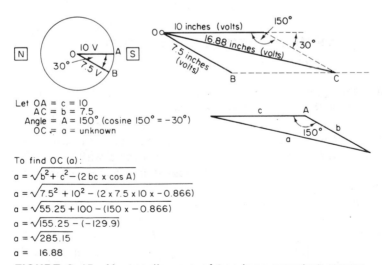

Let OA = c = 10
AC = b = 7.5
Angle = A = 150° (cosine 150° = −30°)
OC = a = unknown

To find OC (a):

$$a = \sqrt{b^2 + c^2 - (2\,bc \times \cos A)}$$

$$a = \sqrt{7.5^2 + 10^2 - (2 \times 7.5 \times 10 \times -0.866)}$$

$$a = \sqrt{55.25 + 100 - (150 \times -0.866)}$$

$$a = \sqrt{155.25 - (-129.9)}$$

$$a = \sqrt{285.15}$$

$$a = 16.88$$

FIGURE 9-15 Vector diagram of two-loop armature generator.

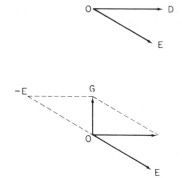

FIGURE 9-16 Vector subtraction by parallelogram.

allel to *OB,* and *BC* equal and parallel to *OA*. Then draw the diagonal *OC*. Measure the length of *OC,* which should be approximately 16.88 in. Line *OC* then represents the *vector sum* of the two voltages, or 16.88 V.

The same problem can be solved using trigonometry, as shown in Fig. 9–15. In this case, a rough vector diagram is laid off to construct a parallelogram. The measurements need not be exact. Then the parallelogram is converted to an oblique triangle as indicated, and the problem is solved by use of the equations shown in Fig. 9–15. The cosines of the angles involved are found in Fig. 9–8.

Vectors can also be used to subtract one out-of-phase value from another, as shown in Fig. 9–16. Assume that *OE* is to be subtracted from *OD*. Reverse the direction of *OE* to get a negative (−) *OE*. Then construct the parallelogram. The resultant line *OG* gives the scale and direction of the difference.

9-10 POWER MEASUREMENT FOR ELECTRICAL DISTRIBUTION SYSTEMS

The apparent power or volt-amperes of an electrical distribution system can be measured with a voltmeter and an ammeter. If the load is inductive (or has any reactance), the apparent power is not the true power. The true power drawn by a load, or all the loads in the system, can be measured directly by means of a *wattmeter*. The basic connections for a wattmeter are shown in Fig. 9–17. These connections are for single-phase power, either pure resistive or inductive.

A typical single-phase wattmeter has four terminals and two circuits. The *current coil* terminals (usually larger in size) are connected in series with the circuit, just as for an ammeter. The *potential coil* or *voltage coil* terminals are connected across the circuit, as in the case of a voltmeter. Note that the ± terminal of the circuit coil is connected to the line side of the circuit, and that

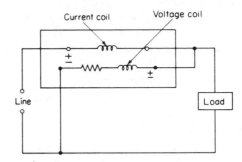

FIGURE 9-17 Basic wattmeter connections for single-phase power measurement.

the ± terminal of the voltage coil is connected to the *same* line lead as the current coil. Reversal of either winding results in a backward deflection.

Wattmeters are designed to recognize power factor angle and to account for the power factor (if any) in their readings. The wattmeter multiplies the current passing through the current coil by the voltage impressed on the voltage coil, and then multiplies this product by the cosine of the angle between the two values. The result is the true wattage of the circuit being measured.

9-10.1 Three-Phase Power Measurement

Several methods are used for measurement of power in three-phase systems. The most common method is shown in Fig. 9–18. This arrangement requires two wattmeters, and applies to three-phase three-wire systems, balanced or unbalanced. The connections of Fig. 9–18 do not apply to three-phase four-

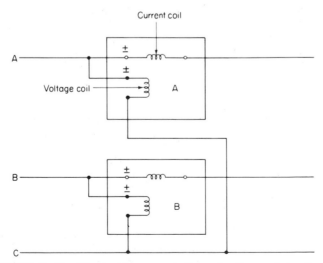

FIGURE 9-18 Wattmeter connections for three-phase three-wire power measurement, balanced or unbalanced loads.

wire systems (wye, with neutral wire), unless the wye system is perfectly balanced. Since the purpose of the neutral in a wye system is to handle possible unbalances in the loads, a special arrangement is used for three-phase four-wire, as discussed in Sec. 9–8.2. Note that in the two-wattmeter method of Fig. 9–18, all common connections are on the line side. Also note that the current of wattmeter A lags the voltage by 30°, and the current of wattmeter B leads the voltage by 90°. (If one wattmeter reads back scale, reverse the current connections.)

If both wattmeters read the same value, the load is balanced and the power factor is 1. The total wattage of the system is found by adding the two wattage readings (wattmeter A + wattmeter B = total system wattage).

If the load is balanced and each wattmeter reads a different value, the power factor is not 1 (there is a combination of inductance and resistance in the loads). The total wattage of the system is found by adding the two wattage readings. The power factor is found from the ratio of the two readings.

The power factor can be found using either the curve or the equation in Fig. 9–19. For example, assume that the wattmeter readings are 240 W and 600 W. Using the curve, the ratio is 0.4 (240/600) and the power factor is 0.8. Using the equation

$$1.73 \times \frac{600 - 240}{600 + 240} = \text{approximately } 0.74$$

the angle with a tangent of 0.74 is approximately 36.5° (Fig. 9–8). The cosine of 36.5° is approximately 0.8 (indicating a power factor of 0.8).

In checking the curve of Fig. 9–19, note that the ratio is 1 for a power

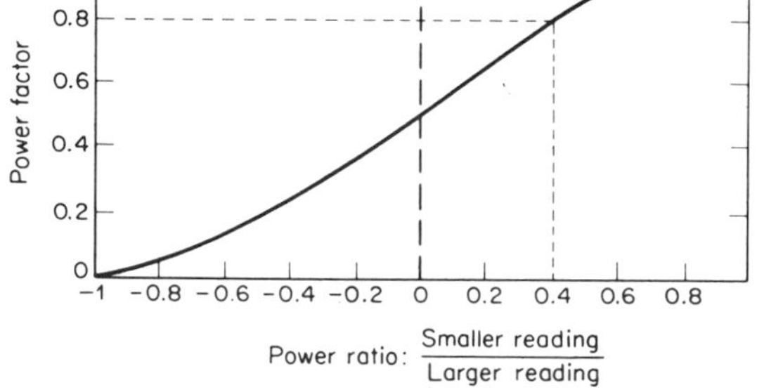

$$\text{Tangent } \theta = 1.73 \frac{W_2 - W_1}{W_2 + W_1}$$

W_1 = Smaller reading

W_2 = Larger reading

Power factor = cosine θ

FIGURE 9–19   Relation between power factor and wattmeter ratio for balanced three-phase loads.

factor of 1 (since the wattmeter readings are equal); the ratio is zero at 0.5 power factor (since the lower-reading wattmeter indicates zero); and the ratios are negative for all lower power factors; and that the ratio is −1.0 at zero power factor (since the wattmeter readings are again equal but the lower reading wattmeter gives a reverse indication).

Keep in mind that the curve and equation of Fig. 9–19 apply only for *balanced loads,* when metered by the two-wattmeter method. Also, if one wattmeter reads zero, the power factor is 0.5. If the power factor is less than 0.5, the angle between one wattmeter current and voltage is greater than 90°. As a result, that particular reading is considered negative and must be subtracted from the other. As a practical matter, the amount to subtract is found by reversing the current connections, so that the wattmeter indicates up-scale rather than back-scale (reverse).

Hints on Practical Wattmeter Measurements. With single-phase loads, a wattmeter deflects backward only if improperly connected. This is not true when the two-wattmeter method is used for three-phase, and the power factor is less than 0.5. When a wattmeter gives a reverse reading, it may be difficult to tell the difference between a low power factor or an improper connection. This can be avoided by paying careful attention to wattmeter polarity markings. The following procedure is recommended for connecting wattmeters for the two-wattmeter method.

1. Connect each wattmeter in the proper manner. That is, connect the ± terminal of the current coil on the line side of the line lead, and

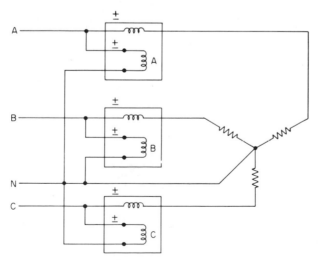

Total system wattage: A + B + C

FIGURE 9-20 Wattmeter connections for three-phase four-wire power measurement, balanced or unbalanced loads.

the ± terminal of the voltage coil to the same line as its own current coil.

2. If either wattmeter deflection is backwards, a low power factor is indicated. Reverse the *current coil* connections and consider this reading as a *negative value*.

9-10.2   Three-Phase Four-Wire Power Measurement

The most common method of measuring three-phase four-wire systems (wye, with a neutral wire for unbalanced loads) is shown in Fig. 9–20. Note that all the voltage coils are connected to the neutral wire. This permits each phase to be measured by a separate wattmeter, and the total power is the sum of their readings (wattmeter A + wattmeter B + wattmeter C = total system wattage).

INDEX